ENVIRONMENTAL RESEARCH ADVANCES

SALT MARSHES

FORMATION, ECOLOGICAL FUNCTIONS AND THREATS

Environmental Research Advances

Additional books in this series can be found on Nova's website under the Series tab.

Additional e-books in this series can be found on Nova's website under the eBooks tab.

ENVIRONMENTAL RESEARCH ADVANCES

SALT MARSHES

FORMATION, ECOLOGICAL FUNCTIONS AND THREATS

DARRIN BARNES
AND
CLAIRE ELLIS
EDITORS

nova
science publishers
New York

Library of Congress Cataloging-in-Publication Data

ISBN: 978-1-53614-040-8
ISSN: 2375-3277

Published by Nova Science Publishers, Inc. † New York

CONTENTS

PREFACE

The main focus of the research described in the opening chapter of *Salt Marshes: Formation, Ecological Functions and Threats* focuses on the study of the plant heritage of the Fuerteventura salt marshes and their surroundings, together with the analysis of their ecological value, threats to the marshes, and of use and recent management practices. The methodology used predominantly consists of field work, photo-interpretation, and implementation of a GIS from the information obtained and the database generated. The Fuerteventura salt marshes are of great importance because of their vegetal wealth and role as a refuge for fauna, especially birds and invertebrates.

The subtropical salt marshes of Brazil are located in the South of Brazil, between the north-central coast of São Paulo state and the central coast of Santa Catarina state, and have been studied over the course of many years by the researchers M. R. Bornschein and B. L. Reinert, resulting in their recognition as a new ecosystem. Despite several ornithological researches in these marshes, little is known about detailed aspects of their ecology and other groups of animals. Thus, the authors provide a general description of this recently recognized ecosystem and the ecological influences over the species that live in these marshes, resulting in the fragile ecological balance point of Borschein-Reinert, and propose a mathematical index to this balance.

Salt marshes develop in estuaries where there is reduced wave action, which allows for a source of sediment and suitable conditions for marsh plants to grow. The vegetation generally provides a structural habitat on featureless soft-sediment bottoms, and so it is utilized by a wide and diverse range of fish and invertebrates as their physical home, food supply and shelter from predators. The authors discuss the way in which species such as insects and crabs demonstrate variable levels of adaptation to life in this intertidal habitat, as well as how species composition and abundance of benthic invertebrates are strongly influenced by changes in the vegetation.

Chapter 1 - The main focus of the research described in this chapter focuses on the study of the plant heritage of the Fuerteventura salt marshes and their surroundings, together with the analysis of their ecological value, of threats to the marshes, and of use and recent management practices. The methodology used consists basically of field work (floristic inventory, study of the most important ecological factors, interviews), photo-interpretation, and implementation of a GIS from the information obtained and the database generated.

The ecological value of the Fuerteventura salt marshes is of great importance because of its vegetal wealth, and its role as a refuge for fauna, especially for birds and invertebrates. It is also relevant because of its vulnerability and uniqueness; this type of threatened ecosystem is under-represented in Macaronesian European region due to the scarcity of flat coastal strips. Thus, salt marshes are exceptional habitats in the Canary Islands where, in addition to the lack of coastal plains, the authors must add the intense environmental degradation of wetlands and coastal areas due to urban and tourism pressures. Despite this, the island of Fuerteventura has the best representation of these ecosystems in the archipelago as a result of its ancient volcanic and sedimentary relief, and the relatively late development of tourism. The main characteristic of these plant formations is their completely fragmented nature and their dispersion along the coast; *Zygophyllo fontanesii – Arthrocnemetum macrostachyi* occupies an area of 87 hectares divided into 70 plots, while *Frankenio capitatae – Suaedetum verae* covers 45 hectares in 40 plots. Finally,

Halimiono portulacoidis – Salicornietum perennis has a unique community of 1,247 m² in the Charco de Bristol.

Chapter 2 - The subtropical salt marshes of Brazil compose a recently recognized ecosystem (only in 2017) and are under human threat. This ecosystem is located in the South of Brazil, between the north-central coast of São Paulo state and the central coast of Santa Catarina state, has less than 20,000 ha and it was studied during many years by the researchers M. R. Bornschein and B. L. Reinert, resulting in being recognized as a new ecosystem. Despite several ornithological researches in these marshes, little is known about detailed aspects of their ecology and other groups of animals. Under strong diary tidal effects and sedimentation processes, although being currently suffering processes of tropicalization, sea level rising and exotic plant species invasions, it is a very dynamic ecosystem. The subtropical salt marshes of Brazil are extremely restrictive ecosystems to animals that live there, being under selection of different factors that result in reduced populations (at least to birds) in these marshes. These animals are frequently under influences of flooding by the tidals, adding up effects of predation, fall of vegetation by wind and influences of extreme climatic events. Therefore, the animal species are in a fragile trade-off to survive, that here the author denominates, in homage to the researchers who first described these characteristics, ecological balance point of Borschein-Reinert. Different from variations in other ecosystem gradients, as latitudinal and altitudinal gradients. Vegetation variations in these marshes, which at first sight appear homogeneous, occur in a few meters. And, for instance, the nest site selection by birds results in strong influences over their reproductive success. The choice of a nest site few meters near or distant the water edge of the marshes or a few centimeters higher or lower, represents the difference between flooding or not flooding of the nest during the high tide. Consequently, the life or death of nestling. Therefore, the species occurring in this ecosystem are living in an extreme environment, always oscillating between environmental characteristics more or less benefic to it, and more successful will be those individuals able to live at these delicate ecological balance. Thus, here the author will discuss a general description of this recently recognized ecosystem and the ecological influences over the species that live in these marshes, resulting in the fragile ecological balance point of Borschein-Reinert, and propose a mathematical index to this balance.

Chapter 3 - Salt marshes develop in estuaries where there is reduced wave action and which is a source of sediment and suitable conditions for marsh plants to grow. The vegetation generally provides a structural habitat on featureless soft-sediment bottoms, and so it is utilized by a wide and diverse range of fish and invertebrates as their physical home, food supply, and shelter from predators. Salt marshes host a unique mix of marine and terrestrial animals including insects, birds, fish, and particularly crabs, whose activities heavily influence the productivity of salt marshes. Most of the primary production and detritus derived from marsh plants are trapped by plants and consumed by terrestrial organisms. Species such as insects and crabs demonstrate variable levels of adaptation to life in this intertidal habitat. Species composition and abundance of benthic invertebrates are strongly influenced by changes in the vegetation. High marshes, made up of shrubs and trees, provide a tridimensional physical structure that supports a considerable biodiversity of such associated animals as insects, small mammals, birds, and reptiles.

In: Salt Marshes
Editors: D. Barnes and C. Ellis
ISBN: 978-1-53614-040-8

Chapter 1

Fuerteventura Salt Marshes (Canary Islands, Spain): Ecological Value, Management and Threats

Salvador Beato Bergua**, *Miguel Ángel Poblete Piedrabuena* and *José Luis Marino Alfonso
Department of Geography, Oviedo University,
Oviedo (Principality of Asturias), Spain

Abstract

The main focus of the research described in this chapter focuses on the study of the plant heritage of the Fuerteventura salt marshes and their surroundings, together with the analysis of their ecological value, of threats to the marshes, and of use and recent management practices. The methodology used consists basically of field work (floristic inventory, study of the most important ecological factors, interviews), photo-interpretation, and implementation of a GIS from the information obtained and the database generated.

* Corresponding Author Email: beatosalvador@uniovi.es.

The ecological value of the Fuerteventura salt marshes is of great importance because of its vegetal wealth, and its role as a refuge for fauna, especially for birds and invertebrates. It is also relevant because of its vulnerability and uniqueness; this type of threatened ecosystem is under-represented in Macaronesian European region due to the scarcity of flat coastal strips. Thus, salt marshes are exceptional habitats in the Canary Islands where, in addition to the lack of coastal plains, we must add the intense environmental degradation of wetlands and coastal areas due to urban and tourism pressures. Despite this, the island of Fuerteventura has the best representation of these ecosystems in the archipelago as a result of its ancient volcanic and sedimentary relief, and the relatively late development of tourism. The main characteristic of these plant formations is their completely fragmented nature and their dispersion along the coast; *Zygophyllo fontanesii – Arthrocnemetum macrostachyi* occupies an area of 87 hectares divided into 70 plots, while *Frankenio capitatae – Suaedetum verae* covers 45 hectares in 40 plots. Finally, *Halimiono portulacoidis – Salicornietum perennis* has a unique community of 1,247 m² in the Charco de Bristol.

Keywords: salt marshes, biodiversity, threats, Fuerteventura, Canary Island

INTRODUCTION

The declaration of Fuerteventura as a Biosphere Reserve in 2009 gives some indication the richness of the island's natural heritage and its state of conservation, with a total to date of 12 protected areas under the figures of Natural Park (3), Natural Monument (6), Rural Park (1) and Protected Landscape (2). It island also has one Site of Scientific Interest (SSI), 9 Special Protection Area (SPAs), and 12 Special Areas of Conservation (SACs).

Research on this heritage has been abundant and widespread, dealing not only with the analysis of the geomorphology (Romero, 1986; Criado, 1991), the volcanic heritage (Dóniz, 2009) and the evolution of the relief (Cubas et al., 1988; Zazo et al., 2008; Criado et al., 2012), but also with biogeographical (Hernández-Cordero et al., 2015), phytosociological (Rodríguez et al., 2000), and botanical (Kunkel, 1977; Santana & Naranjo,

2002) aspects, as well as with the historical evolution of the plant landscape (Criado, 1990), and the whole natural heritage of the island (Rodríguez, 2005). More specific studies of various protected natural areas have also been carried out, especially on the Dunas de Corralejo e Isla de Lobos Natural Park (Criado et al., 2004; Criado et al., 2007; Fernández et al., 2012).

Salt marshes are exceptional habitats in the Canary Islands given the lack of coastal plains. However, they have suffered intense environmental degradation as a result of urban and tourism pressures, which has endangered the high ecological value of these wetlands. Even so, the island of Fuerteventura has the best representation of these ecosystems in the Canaries archipelago due to its ancient volcanic relief, the different processes of modeling and sedimentation, and the late development of tourism. Despite the low rainfall, high insolation, strong winds, high salinity, and restrictive soils, the salt marshes on Fuerteventura display enormous diversity and rich vegetal wealth on being made up of hygrophilic, halophilic, xerophilous and psammophilic communities, with the ecosystems of thermo-Atlantic halophilic scrub proving especially noteworthy.

In spite of their uniqueness within the Canaries archipelago, not all of the salt marshes of Fuerteventura are protected. However, the Playa del Matorral, also known as the Saladar de Jandía, has been the subject of several studies aimed at its declaration as a SSI and SAC, and at the recovery of its outstanding natural values. Not in vain is it the largest salt marsh in the archipelago and one of the most important in the Macaronesian region. Some eight kilometers further north along the coast is the Playa de Sotavento salt marsh, which is included in the Jandía SAC (which borders the Playas de Sotavento de Jandía marine SAC). As a result, it also figures in environmental catalogues and in the relevant management documents, although somewhat sparingly given the territorial and ecological scope of the SAC. The same is true of the small pockets of salt marsh on the *malpaís* (aa lava flows) of Jacomar-Toneles, included in the Pozo Negro SAC.

For its part, the Charco de Bristol has gone largely unnoticed, and due to its reduced dimensions has only been the subject of one specific investigation (Beato et al., 2017). In addition, it is cited in botanical research on specific species (Fernández & Santos, 1983) and in more general papers on the vegetation of the island and the Canaries archipelago (Del Arco et al., 2006). These papers evidence its uniqueness as the only island enclave in which a population of *Sarcocornia perennis* (community *Halimiono portulacoidis – Salicornietum perennis*) appears. Despite this, the Charco de Bristol has not been granted any form of protection.

These coastal salt marshes constitute a plant heritage of extraordinary value in the Canary Islands, and one that needs to be preserved. To this end, it is essential to recognize the relevant ecological function that they play, to determine their current conservation status, and to identify the threats that may be hovering over them, all with the aim of applying restoration and protection measures that will serve as the basis for adequate environmental management in the future.

Study Area

The island of Fuerteventura is the second largest of the Canaries archipelago with an area of 1,659.74 km², and, along with Lanzarote, is one of the easternmost. Situated less than 100 km from Africa, the island lies between 28º45'25.64" and 28º02'38" North, and 14º30'35.84" and 13º49'31.00" West (Figure 1). In general, Fuerteventura is relatively flat and low lying (most of the island does not exceed 500 m a.s.l.), especially in the northern and central areas. These are only broken by old volcanic massifs such as Betancuria, Tetir and Jandía, where the Pico de la Zarza is located. At 807 m a.s.l. this is the highest point on the island. From the physiographic point of view, from N to S we have the following units: the northeast coastal plain, the central plain, the valleys and crest-shaped interfluves of the eastern sector, the Betancuria massif, and the peninsula of Jandía (Criado, 2005).

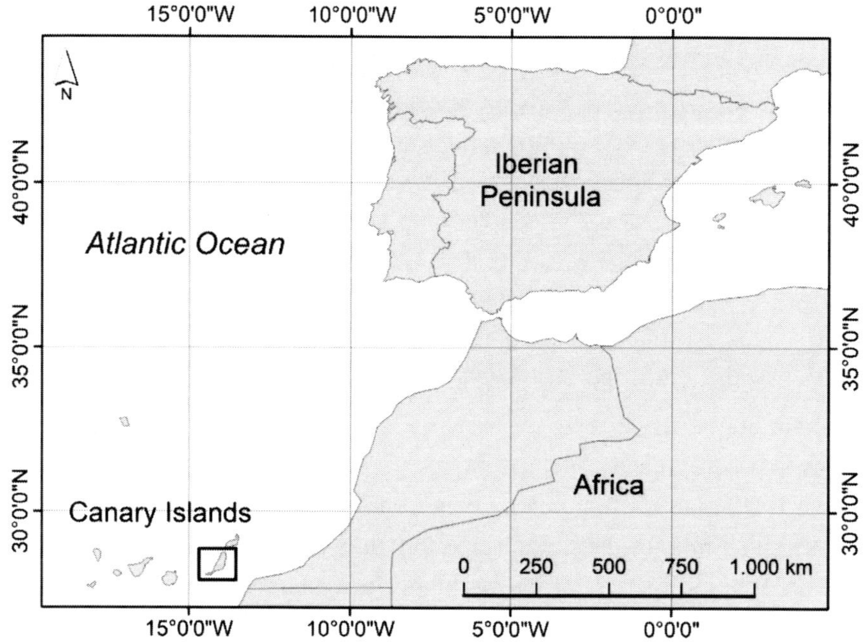

Figure 1. Location of Fuerteventura in the Canary Islands.

The island of Fuerteventura, like the rest of the archipelago, is volcanic in origin and has been formed from numerous eruptions, the oldest of which are more than 20 million years old. Quaternary and Holocene volcanic forms are scarce in general, with erosive morphogenetic processes predominating under arid morphoclimatic conditions. However, the extreme north of the island is the result of Quaternary volcanism that corresponds to Volcanic Series III and IV in the so-called Basament Post-Complex (Fúster et al., 1968; Ibarrola et al., 1989; Coello et al., 1992; Ancochea et al., 1993). This basaltic composition volcanism stands out morphologically because of the presence of large cinder cones, some of which are the largest on the island (Series III), and because of the configuration of a Würmiense age (Series IV) volcanic chain of isolated and imbricated cones concentrated in the central and northern sectors of the island. This activity took place between 51,000 and 26,000 B.P. (Pomel et al., 1985; Criado, 1991). The rest of the island is formed, in general, from ancient basaltic Miocene massifs where the most characteristic morphological feature is the modeling or excavation of wide U-shaped valleys.

The climate Fuerteventura enjoys is determined by three basic factors: the proximity of the continent of African, the predominance of the NE trade winds, and the cold ocean current that runs along its coasts (Marzol, 1993). These factors mean that subsidence inversions which produce marked atmospheric stability are very frequent, with the subtropical anticyclone of the Azores being chiefly responsible for this situation (Dorta, 1996). Thus, the most defining climatic feature of Fuerteventura is, without doubt, its extreme dryness; the data provided by the AEMET (the State Meteorological Agency) at the El Matorral Airport for the series 1981-2010, shows an annual average rainfall of 97 mm spread over 15.7 days, which implies that it rains no more than 4.3% of the year. In addition, the distribution of this rainfall is highly irregular, with more than half of the rains concentrated in winter (57.7%), followed by autumn (23.7%) and spring (18.6%), while in summer there is no precipitation. If we add to this irregular distribution the fact that the rains are frequently torrential in nature, we can see that they constitute a clear restrictive factor for the development of vegetation and edaphic formations, since they trigger strongly erosive runoffs.

Temperatures on Fuerteventura are very mild, with an annual average of 21.1ºC. January is the coldest month of the year, with an average of 17.6ºC, and August the hottest, with an average of 24.5ºC. Thus, at only 6.9°C, the thermal amplitude is small, whilst the average relative humidity is 70% thanks to the influence of the ocean. To this we have to add a high average insolation of 2,836 hours of sunshine per year, together with the strong winds that batter the island all year round. These favour evaporation, influence the configuration of any vegetation, and act as an erosive agent. In general, northerly winds predominate (Criado, 1991) with average speeds between 7 and 10 m/s (Ministerio de Medio Ambiente, 2006). Unlike with the western islands, the trade winds have no significant bioclimatic effects on Fuerteventura due to its low relief. As a result, there are hardly any differences between windward and leeward slopes.

Finally, the plant landscape of the island of Fuerteventura, deeply transformed by human activity, is characterized by a clear predominance of substitution shrublands, the most representative potential natural formations of the island being limited to small coastal enclaves and tiny patches in the Betancuria and Jandía massifs. These plant communities include: *Lycio intricati – Euphorbietum balsamiferae*, *Olea cerasiformis* woodlands (*Micromerio rupestris – Oleatum cerasiformis*), edaphohaline small tree vegetation (*Suaedo verae – Tamaricetum canariensis*), *Euphorbia* scrubs (*Euphorbietum handiensis* and *Kleinio neriifoliae – Euphorbietum canariensis*), and Canary palm community (*Periploco laevigatae – Phoenicetum canariensis*).

Materials and Methods

The methodology consisted basically in field work and, to a lesser extent, in the photointerpretation of aerial images (years 1981, 1984 and 1989) and the orthophoto map of the National Plan of Aerial Orthophotography. In addition, digitized information was consulted from the Canary Islands Spatial Data Infrastructure and the National Center of Geographic Information (Spain). Despite their usefulness when performing geomorphological cartography and when distinguishing different vegetation units (Hernández-Cordero et al., 2015), aerial images have proven less useful in the recognition and identification of certain salt marsh vegetal communities because of the reduced dimensions, and, above all, the scarce imprint shown by this low-size, low-density vegetation. For this reason, our fieldwork has been exhaustive, with all of the biotopes analyzed through multiple transects carried out between October 2013 and December 2017.

First, sketches were made on the orthophoto maps of the National Area Orthophotography Plan (corresponding to the 2012 photogrammetric flight) with the identification, a priori, of the surfaces occupied by the main vegetal formations according to texture, color and the spatial distribution. The verification of the limits of the vegetation units and the in depth phytosociological study were undertaken by two methods. On the one hand, different linear routes were designed to make a stratified random sample from one meter wide transects of various lengths. On the other, a number of 2 x 2 m plots were designated in the most representative geomorphological units (open coast, lagoon, delta, restricted-entrance embayment, low coast of aa lava flows, low cliff with gullies and lower courses, and mouths of deep gorges) in order to carry out a more thorough herborization. With the aim of characterizing, interpreting, and mapping the different plant communities (Beltrán et al., 1999; Arozena, 2000; Hernández-Cordero et al., 2015), a floristic study was carried out; vascular plants were inventoried by strata, vegetation units were delimited based on abundance/dominance criteria (Braun-Blanquet, 1979), and the most relevant ecological factors were analyzed. After this, an assessment of the state of conservation was carried out, together with a typification of the hazards that threaten the vegetation. The evolution of uses was studied and recent impacts were identified, in particular, from the transit of people and vehicles, as well as from any uncontrolled discharges made. Finally, we interviewed visitors to the salt marshes, construction entrepreneurs, and officials from local councils and the Cabildo de Fuerteventura (the Regional Government).

With the information obtained, a database included in a GIS was developed using the ArcMap 10.1 software (© ESRI), the polygons corresponding to the distribution of the different plant communities were digitized, and the surface calculations were obtained together with the definitive cartographic representation. In this way, the foundations were laid for future monitoring of the evolution of the salt marshes in Fuerteventura similar to that already in place in the Charco de Bristol (Beato et al., 2017), the dunes of Maspalomas, and the island of La Graciosa (Hernández et al., 2014; García-Romero et al., 2014).

FORMATIONS AND ECOLOGICAL VALUES OF THE SALT MARSHES IN FUERTEVENTURA

Salt marshes lie on the margins of coastal areas that are formed by fine sediments (sands and mud), covered by halophytic vegetation, and flooded periodically by sea water thanks to the ebb and flow of the tide. They therefore configure complex ecosystems which are transitional –terrestrial and maritime– and dynamic –they are constantly changing and evolving– and which play an important role in many coastal environments. Their most relevant functions are geomorphological –they protect the lowlands from sea floods through the attenuation of waves and the reduction of processes of coastal erosion– and ecological –they control pollution and water quality, provide nutrients for coastal waters, and offer varied habitats for wildlife (spawning fish, birds, reptiles, crustaceans, insects, etc.)– (Allen, 2000; Adnitt et al., 2007; Doody, 2008; Mcowen et al., 2017).

The salt marshes of Fuerteventura, especially those located on the Jandía peninsula, are of great floristic richness and high ecological quality. They serve as a reproductive support, shelter and trophic reservoir for a great variety of terrestrial fauna, from endemic invertebrates, through reptiles, to nesting waders and migratory sea-birds. In fact, more than 40 taxa associated with endemic humid environments have been inventoried – 2 flora and 39 invertebrate– with 16 species endemic to the Fuerteventura salt marshes and 23 that live in their surroundings (Díaz, 2007). The endemic *Gallotia atlantica mahoratae* lizard stands out as the most unique species. Of note among the many nesting birds are the endemic *Saxicola dacotiae* and *Anthus berthelotii*, whilst among bird sightings there is the endangered endemic subspecies *Chlamidotis undulata fuerteventurae*. Finally, mention must be made of the leatherback sea turtle (*Dermochelys coriacea*), which lays its eggs in these beaches, the only reproductive site of the species in Europe (Díaz, 2007).

The salt marshes of Fuerteventura belong to the low infra-Mediterranean, low arid bioclimatic belt. They are formed by communities of hygrophilous, halophilic, xerophilous and psammophilic plants that are biogeographically located in the Majorero Sector of the Eastern Canarian Subprovince (Rivas-Martínez et al., 2002). Their most significant plant wealth is integrated into the thermo-Atlantic halophilous scrub ecosystem. The habitat is characterized by the class *Salicornietea fruticosae* Br.-Bl. et Tx. ex A. Bolòs y Vayreda et O. de Bolòs in A. Bolòs y Vayreda 1950, in which chenopodiaceous thickets predominate made up of saline soil, perennial, woody formations (Rivas-Martínez et al., 2002; Del Arco et al., 2006; Del Arco et al., 2010).

Although the potential total surface area for all of the Canarian salt marshes is estimated to be 362 ha, only 304 ha remain (Del Arco et al., 2010) because of their being located in coastal areas subject to strong anthropic pressure. In Fuerteventura, the salt marshes are scattered along the coast. *Zygophyllo fontanesii – Arthrocnemetum macrostachyi*, for example, occupies an area of 87 hectares divided into 70 different plots, while *Frankenio capitatae – Suaedetum verae* covers 45 ha in 40 plots. Finally, a population of *Sarcocornia perennis* (community *Halimiono portulacoidis – Salicornietum perennis*) occupies less than 0.2 ha in 1 plot at the Charco de Bristol (Corralejo). Thus, the salt marshes of Fuerteventura occupy a total area of 132 ha (Del Arco et al., 2010), which represents 43.4% of the total of the Canary Islands salt marshes, whose vegetation belongs to the class *Sarcocornietea fruticosae* (Figure 2).

Arthrocnemum macrostachyum resists partial flooding during high tide. It presents stems that are woody at the base but articulated and fleshy above, and it spreads out in the form of scattered bushes up to 1.50 m in height. It is a generally upright shrub, although it sometimes appears prostrate. In the outermost range of the salt marshes away from the daily tides, it has to withstand long periods of drought and the strong competition of *Suaeda vera*.

The role of *Arthrocnemum macrostachyum* in biological interactions is very interesting since it modifies physical-chemical aspects of the soil (Sadek & Eldarier, 1995; Pugnaire et al., 2004; Espinar, 2009) facilitating the colonization of other species. In addition, the species is included in the 2010 Canary Islands Catalogue of Protected Species, where it is classed as being of special interest for Canarian ecosystems. In the 2009 assessment commissioned by the Canarian Government, despite appearing to present a degree of stability, it was felt to show a regressive tendency based on data from previous decades, and its location in areas of urban development (Rodríguez & Valdivia, 2009).

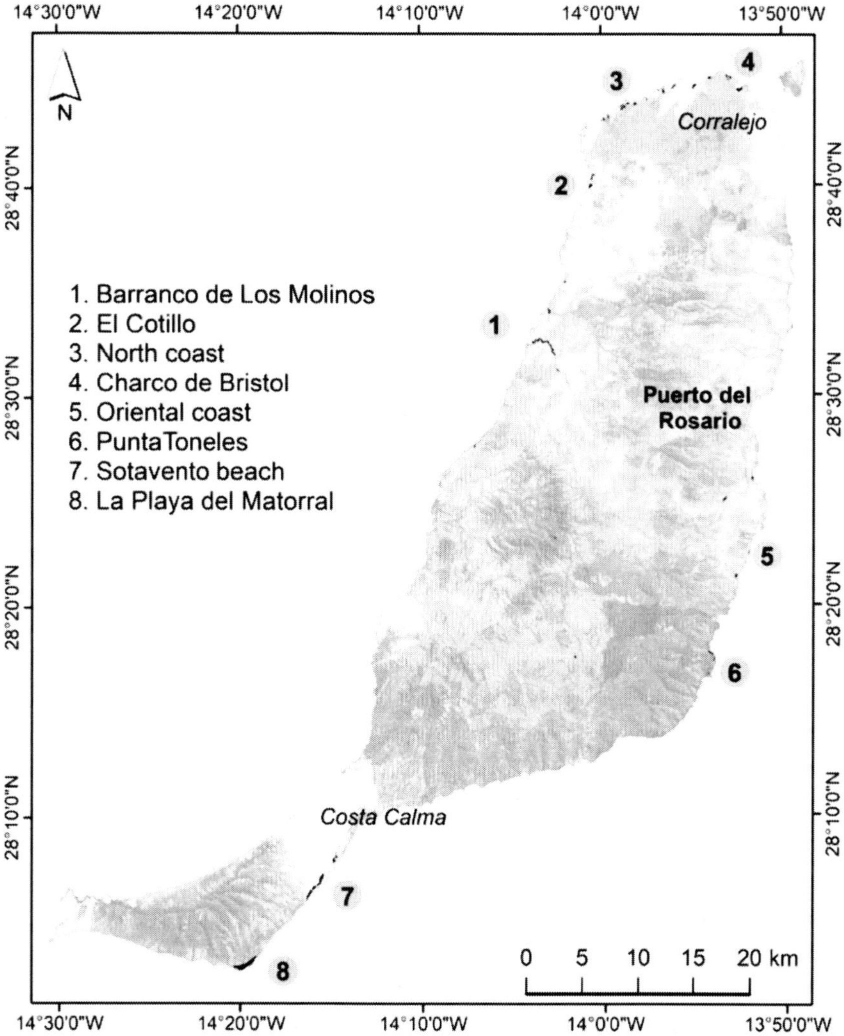

Figure 2. *Sarcocornietea fruticosae* in Fuerteventura (own elaboration based on information supplied by Del Arco et al., 2006).

Sarcocornietum perennis (Fernández & Santos, 1983), name of the plant association recently corrected to *Halimiono portulacoidis – Salicornietum perennis* by Del Arco et al., (2017), is also defined in the European Union Habitats Directive as being of Community interest (Espinar, 2009). This plant community is located in the transition from the foreshore to the backshore of the salt marsh, that is to say, at the lower level where, twice a day, it resists partial or total flooding at high tide. It thrives, therefore, on strongly silty materials deposited by the sea, is characterized by a continuous carpet or lawn. Strongly branched, it has a woody base and fleshy articulated stems in the upper part, where salts accumulate. In addition to the single community of *Sarcocornia perennis* on the island of Fuerteventura, there are only three other enclaves in the eastern Canary Islands (Del Arco et al., 2006), because of which it is included in the Canary Islands Catalogue of Endangered Species as a species of high interest (Figure 3).

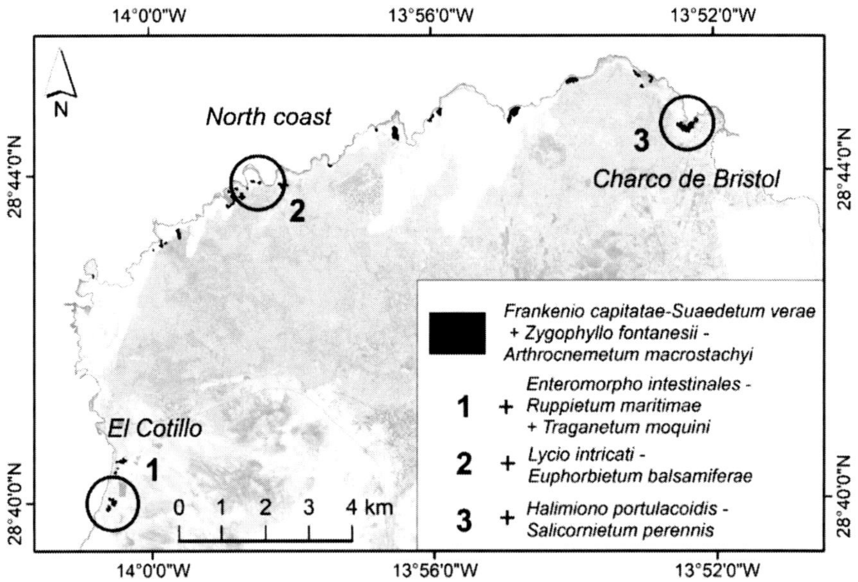

Figure 3. Most relevant vegetal formations in the salt marshes of the north of Fuerteventura (own elaboration based on information supplied by Del Arco et al., 2006).

There are other hemicriptophites and camephites that are characteristic of the plant associations of the island's coastline, such as *Tetraena fontanesii* and *Frankenia capitata*, of the Zygophyllaceae and Frankeniaceae families, respectively. Vegetation from the class *Ruppietea maritimae* (*Enteromorpho intestinalis – Ruppietum maritimae*) is also present in an enclave of El Cotillo, as is the *Lycio intricati – Euphorbietum balsamiferae* series in various localities along the north coast.

The main plant formations in these brackish environments are accompanied by others that correspond to substitutional shrublands as a result of the intense anthropization suffered by some of these coastal enclaves. In particular, plants of the regressive series *Chenoleoideo tomentosae – Suaedetum molli* and *Chenoleoideo tomentosae – Salsoletum vermiculatae* are recurrent.

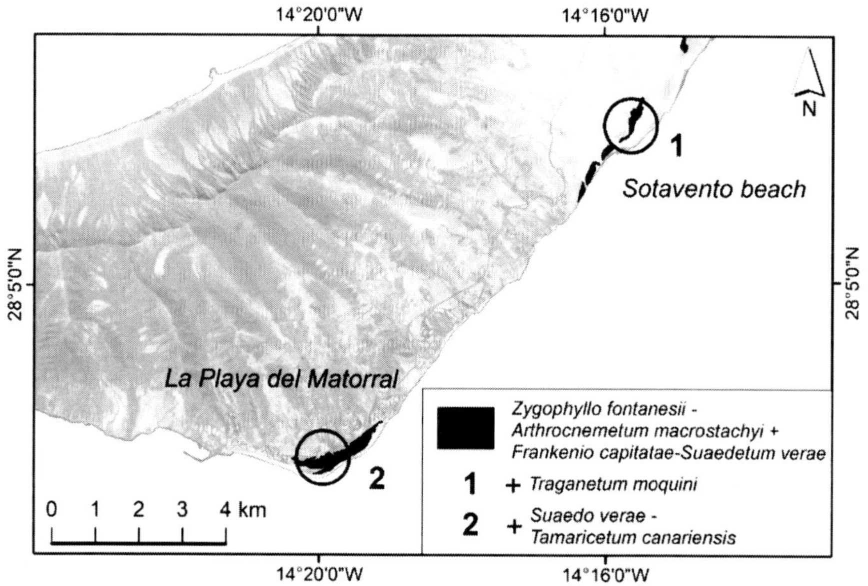

Figure 4. Principal vegetal formations in the salt marshes of the south of Fuerteventura (own elaboration based on information supplied by Del Arco et al., 2006).

In areas where sands abound, there are also taxa of *Polycarpaeo niveae – Lotetum lancerottensis*, especially in the northern areas. In addition, the beach salt marshes are enriched both with vascular vegetation of the association *Traganetum moquini*, especially where the sands are more abundant or even form dunes. *Suaedo verae – Tamaricetum canariensis* can also be found in the southern enclaves (Figure 4).

Table 1. Geomorphological types of salt marshes in Fuerteventura

Name	Area (ha)	Geomorphological setting	Dominant vegetation	Other vegetation
Barranco de Los Molinos	11.82	Gorge with rambla	*Frankenio capitatae-Suaedetum verae*	*Chenoleoideo tomentosae-Suaedetum mollis*
El Cotillo	4.03	Low cliff with gullies	*Zygophyllo fontanesii-Arthrocnemetum macrostachyi*	*Frankenio capitatae-Suaedetum verae* *Entheromorpho intestinalis-Ruppietum maritimae* *Traganetum moquini* *Polycarpaeo niveae-Lotetum lancerottensis* *Chenoleoideo tomentosae-Suaedetum mollis*
Playa de Sotavento	30.06	Lagoon and open coast	*Zygophyllo fontanesii-Arthrocnemetum macrostachyi*	*Traganetum moquini* *Polycarpaeo niveae-Lotetum lancerottensis* *Chenoleoideo tomentosae-Suaedetum mollis*
North coast	21.34	Low coast of aa lava flows Restricted-entrance embayments	*Frankenio capitatae-Suaedetum verae*	*Zygophyllo fontanesii-Arthrocnemetum macrostachyi* *Chenoleoideo tomentosae-Suaedetum mollis* *Lycio intricati-Euphorbietum balsamiferae* *Polycarpaeo niveae-Lotetum lancerottensis*
Charco de Bristol	5.78	Restricted-entrance embayment	*Frankenio capitatae-Suaedetum verae*	*Zygophyllo fontanesii-Arthrocnemetum macrostachyi* *Halimiono portulacoidis – Salicornietum perennis* *Polycarpaeo niveae-Lotetum lancerottensis* *Chenoleoideo tomentosae-Suaedetum mollis*

Name	**Area (ha)**	**Geomorphological setting**	**Dominant vegetation**	**Other vegetation**
Oriental coast	6.94	Gorges	*Frankenio capitatae-Suaedetum verae*	*Chenoleoideo tomentosae-Suaedetum mollis*
Punta Toneles	3.86	Low coast of aa lava flows	*Zygophyllo fontanesii-Arthrocnemetum macrostachyi*	*Frankenio capitatae-Suaedetum verae* *Chenoleoideo tomentosae-Suaedetum mollis*
La Playa del Matorral	46.92	Delta	*Zygophyllo fontanesii-Arthrocnemetum macrostachyi*	*Frankenio capitatae-Suaedetum verae* *Suaedo verae-Tamaricetum canariensis* *Polycarpaeo niveae-Lotetum lancerottensis* *Chenoleoideo tomentosae-Suaedetum mollis*

Based on the geomorphological classifications of salt marshes elaborated by Dijkema (1987), Pye and French (1993), Curtis and Sheehy Skeffington (1998), and Allen (2000), we have distinguished seven types of salt marshes in Fuerteventura: open coast, lagoon, delta, restricted-entrance embayment, low coast of aa lava flows, low cliff with gullies and lower courses, and mouths of gorges (Table 1). In all the coastal forms mentioned above, the salt marshes are characterized by a wide biodiversity of communities and, in particular, by the predominance of the formation *Zygophyllo fontanesii – Arthrocnemetum macrostachyi*. In contrast, in the mouths and the lower reaches of deep gorges such as those of Los Molinos, La Muley, La Cueva and Puerto de La Torre, distant from the influence of the sea but through which brackish waters circulate seasonally, the salt marshes are uniform and are formed exclusively by the formation *Frankenio capitatae – Suaedetum vera*. Finally, in the clay deposits located behind the large, sandy beaches in the south of the island (Playa del Matorral and Playa de Sotavento) *Arthrocnemum macrostachyum* dominates over large areas despite its coexistence with taxa from other series such as *Frankenio capitatae – Suaedetum verae*.

The heritage value of *Sarcocornietea fruticosae* is reflected in its cataloguing as a European Union (EU) habitat of Community interest under code 1420. The Habitats Directive (92/43 / EC) of the EU created Natura 2000 in 1992 to serve as a tool for the conservation of natural habitats and wild fauna and flora in Community territory. Consequently, the habitats in which *Sarcocornietea fruticosae* participates justify the official declaration of protected natural areas. A clear example of this is found in the ecological link between salt marshes and birds: in Fuerteventura all the areas where habitat 1420 is mapped are located within a SPA, except for some enclaves adjacent to urban spaces.

The Northern Salt Marshes

On the north coast of Fuerteventura there are small scattered salt marshes mostly dominated by the association *Frankenio capitatae – Suaedetum verae*, with extensions between 1 and 6 ha. Shrubby sea-blite (*Suaeda vera*) has a fundamental role in shaping practically monospecific communities, either in the form of ample and dense herbaceous and sub-bushy size carpets, or in the form of isolated bushes on clays or sands, and even on the *callaos* (deposits of cobbles and boulders rounded or flattened by wave action) located between the Faro del Toston and Corralejo. A good example of the adaptability of this species is in Caleta de Beatriz, where *Suaeda vera* extends through dunes, clay depressions, and cobble deposits. Thus, provided that the geo-ecological conditions of salinity, humidity and nitrification are met, it participates in several associations found in the north of Fuerteventura and other saline enclaves of the island, such as in *Zygophyllo fontanesii – Arthrocnemetum macrostachyi* and *Frankenio ericifoliae – Zygophyllum fontanesii*, or as in the regressive formations of *Chenoleoideo tomentosae – Salsoletum vermiculatae* and *Chenoleoideo tomentosae – Suadetum mollis*. On sands, it coexists with *Polycarpaeo niveae – Lotetum lancerottensis*. In addition, between the Beatriz and the Rincón coves it forms part of several communities mixed with *Lycio intricati – Euphorbietum balsamiferae*, the association endemic to the island.

Suaeda vera can withstand sporadic flooding but remains in the outer zone of the salt marshes, above the foreshore. This happens in the Majanicho salt marsh, a 3.5 ha coastal depression where the *malpaís* is covered with clays, and where *Arthrocnemum macrostachyum* only questions the domain of shrubby sea-blite in the more depressed, and therefore occasionally flooded, central zone. Although some examples of *Suaeda vera* appear in isolation, they generally occur in dense bushes that attain an important level of cover, and scarcely leave space for other species (*Frankenia laevis* var. *capitata*). One exception to this is when it appears next to *Arthrocnemum macrostachyum* in the center of the salt marsh, or together with *Launaea arborescens* and *Salsola vermiculata* on its margins (Figure 5).

Figure 5. The Majanicho salt marsh at the North coast.

Zygophyllo fontanesii – Arthrocnemetum macrostachyi is only dominant in two small enclaves on the north coast between Bajo San Juan and Punta Brava, although always in coexistence with *Frankenio capitatae – Suaedetum verae*. The Charco de Bristol salt marsh is of note since it hosts the island's only population of *Sarcocornia perennis*, and is a perfect example the ecological functions of this type of space, and of the mismanagement and the threats that loom over them.

The most characteristic morphostructural feature of the northern relief of Fuerteventura is the Montaña Colorada-Bayuyo alignment, composed of nine volcanic cones lying along a 5 km, NE-SW structural axis (Criado, 1991). The basaltic lava flows from these volcanoes, especially from Bayuyo, make up a large *malpaís*. This not only increases the island's surface by 110 km^2, but also produces important interferences along the coast, fossilizing a Paleocosta with marine deposits and an old *jable* (sand deposit) (Criado, 1991). The Holocene marine processes model a low-rise cliff on this *malpaís* in this sector. More exactly, the salt marshes lie on depressions dug into this active cantile, and wherever the lavas come into contact with the sea. From a morphological point of view, these lavas present a scoriaceous, highly chaotic external appearance, with small elevations of rocks laid bare by erosion alternating with depressed or subhorizontal areas where fine materials accumulate. In fact, the *malpaís* is covered in many places by bioclastic sands of remobilized marine origin, especially where the topography and the dominant wind direction allows. These are Pleistocene and Holocene sands with clays and Saharan dust that form discontinuous wind mantles which, in many cases, completely cover the basalt lavas.

Climatic conditions are one of the main factors that influence the spatial distribution of plant formations, together with their physiognomic and even floristic features. These characteristics are similar over a good part of Fuerteventura and have been commented on earlier. Nevertheless, it is worth remembering the scarcity of rainfall, which is even more pronounced along this northern coastal strip, where it is less than 100 mm per year (Marzol, 1988), and where evapotranspiration rates can exceed 800 mm per year (Mora et al., 2009). Added to this are high levels of insolation and mild temperatures, to which we further need to add the fierce winds and high salinity of these coastal areas. That is to say, we are looking at an environment that is so restrictive that it even complicates the development of halophilic and xerophilic vegetation adapted to environments subjected to severe water stress. In short, the climate of the northern coastal rim of Fuerteventura is type BWhsi', that is to say, semi-isotherm warm desert (López & López, 1979). To these regional climatic characteristics we need to add other constraints of a local nature that also affect the distribution of vegetation. This is the case of the topography, which basically interferes in the force and direction of the winds, as well as of the environmental humidity coming from the sea.

The soil is the other basic factor that explains the arrangement of the halophyte vegetation of the northern salt marshes, four of the main types being present on the island. All of them are characterized by an aridic moisture regime. In particular, in the central areas of the salt marshes there are salids soils which contain a great amount of soluble salts, and which come up from the water table because of the evaporation of surface water. They have a silty texture and low stoniness, although they can present pebbles and transported boulders on the surface. On the other hand, the dune spaces have psamment-type entisols, that is, sandy soils, in this case of marine bioclastic origin, without differentiated horizons. They also show a degree of salinity due to the influence of the sea, and in some cases are nitrified by urban pollution and the presence of decomposing marine plant remains. The rest of the area is occupied by basaltic lavas without soil development and orthents, namely stony soils of little or no thickness and with saline content, sometimes carbonated, associated with others a little more evolved (calcids). According to the FAO/UNESCO classification, these soils are solonchaks (salines), arenosols (sandy) and leptosols (rocky outcrops). In addition there are anthrosols in the areas used for small constructions or as waste dumps during the last decades of the 20th century.

The profound transformation that the vegetal landscape of Fuerteventura has suffered (Criado, 1991) explains the current huge expansion shown by ruderal associations. Thus, in the salt marshes of the north different forms of substitution shrublands can be found; on sandy soils that are more or less compacted and under constant aerohaline influence, *Chenoleoideo tomentosae – Salsoletum vermiculatae* has developed where *Salsola vermiculata, Chenoleoides tormentosa*, and *Lycium intricatum* participate in the immobilization of the sands. *Suaeda mollis* appears instead of *Salsola vermiculata*, forming the association *Chenoleoideo tomentosae – Suaedetum molli*. These nitrophilous communities have spread practically throughout the island due to anthropogenic degradation, and in the surroundings of the salt marshes they gain importance inland, with *Launaea arborescens* present to a significant degree.

Additionally, as we have already mentioned, the wide ecological amplitude of *Suaeda vera* guarantees important extensions in almost all of the plant communities, including substitution communitites like arid nitrophilous scrub. *Tetraena fontanesii* or *Frankenia laevis* var. *capitata* participate or coexist with these regressive plant associations always under the influence of aerohaline. In the most anthropized spaces, the presence of the nitrophilic scrub dominated by *Nicotiana glauca* is also noteworthy.

The Charco de Bristol Salt Marsh

The Charco de Bristol on the north of the island of Fuerteventura next to the town of Corralejo, is a small coastal depression in the *malpaís* from the Bayuyo volcano emissions. To the NE, a seawater inlet traverses the Pleistocene basalts and partially floods the area during high tide, continuously providing the marine sediments and mineral salts that condition the plant formations. In addition, salinity is higher because of splashing waves, marine spray, and capillarity. All of this contrasts starkly with the extreme scarcity of freshwater entering the Charco de Bristol, both because of the island's very low rainfall, and because of the lack of fluvial channels.

The location of the marsh, protected by currents and winds from the NW, also promotes the accumulation of sands, silts and clays in the most sheltered section, where the geo-environmental conditions are favorable. To the E of the marsh there is also a small dune system fixed by vegetation and bounded by a decorative cover of scories and pyroclasts that links to the urbanized space. On the seaward side, in the section N of the salt marsh, there is a *callao* several meters wide and more than two meters high. In addition to shingle, it is made up of materials of organic origin such as calcareous algae and shells. Finally, a marine abrasion platform in which the Pleistocene basalts appear to be cut at a bevel, extends beneath the *callao*.

The richness and the diversity of the Charco de Bristol and its surroundings, with the presence of numerous biotopes, are determined by several factors. Among these are the landforms, the characteristics of the substratum, the distance from the coast, and the processes of wind deflation and sedimentation (Hernández-Cordero et al., 2015). Thus, to the west, coastal halophylic rock scrub (*Frankenio ericifoliae – Zygophylletum fontanesii*) develops on the upper part of the *callaos* along the entire shoreline and on the rear edge. This vegetal association, which is preserved to a large extent in the rocky belt of the arid and hyper-arid coasts of the eastern Canary Islands and Tenerife (Del Arco et al., 2006; Del Arco et al., 2010), covers an area of 404 ha on the island of Fuerteventura (Del Arco et al., 2010).

This coastal rock association, adapted to strong winds and saltpetre, has a strong presence in the Charco de Bristol. It develops in places where it can get the salt contributions it needs directly from the sea without being flooded, and also in lithic, leptosol-like soils that are heavily laden with salts (Mora et al., 2009). Thus, it prefers the boulders and cobbles deposited right on the shoreline, since they are raised up above the level of the high tide and lie on rocky substrates that drain water easily. The community *Frankenio capitatae – Zygophyllum fontanesii* does not form dense patches of vegetation, and is limited to more or less isolated, scattered individuals.

In the eastern part of the Charco de Bristol salt marsh, the characteristic species of this association (*Tetraena fontanesii* and *Frankenia laevis* var. *capitata*) are located on the cobbles and boulders that come between the sea and the dune space. Their appearance is sparse and scattered, with bushes of *Suaeda vera* intruding increasingly. In this sense, Rodríguez et al., (2000) point to the constitution of a new sub-association endemic to Fuerteventura, *Frankenio capitatae – Zygophylletum fontanesii suaedetosum verae*. The dominance of the shrubby sea-blite coincides with that produced to the north of the study area, and in general with that produced in a good part of the north coast of Fuerteventura (Figure 6).

Figure 6. The Charco de Bristol salt marsh at Corralejo (La Oliva).

Limonium papillatum enriches this community and grows in the western section of the Charco de Bristol under the influence of the prevailing winds and marine spray, growing both on the *callao* and on sandy and rocky substrates. Known locally as *siempreviva zigzag*, *Limonium papillatum* is a Macaronesian endemism included in the Canary Islands Catalogue of Protected Species due to its fragility in the face of habitat alteration. It is a very characteristic plant, small in size (between 10 and 15 cm in height), with prostrate branches and zigzagging, papillose inflorescences with pink and whitish florets. It is scattered around, with some isolated individuals being found on the *callao*, and to a lesser extent, behind it, where it participates in other associations, such as those of substitution scrubs.

The communities typical of the salt marsh, the associations *Zygophyllo fontanesii – Arthrocnemetum macrostachyi* and *Frankenio capitatae – Suadetum verae*, are found behind the sandy-clayey covering.

An accumulation of basaltic lavas about three meters high protects the depression that houses the salt marsh, and despite being practically devoid of vegetation due to the lack of soil and soil moisture, it is covered with lichens, examples of *Suaeda vera*, and species native to arid nitrophilous substitution scrub. The basalts are crossed by a natural channel more than 100 m long. This channel allows the sea to break the abrasion platform carved on the basalts and so to flood the small depression of the Charco de Bristol. This entrance of marine water opens out towards the west, where the weaker influence of winds and waves facilitates the deposition of muds in the intramareal space, and the development of *Sarcocornia perennis*.

Morales (2009) estimated that *Sarcocornia perennis* was distributed over four 500 m UTM squares in Fuerteventura and on the Isla de Lobos, totalling an area of 1.75 km^2. In a follow-up investigation, Santana and Naranjo (2002) indicated that it had an area of 4,000 m^2 in the Charco de Bristol, the only enclave on Fuerteventura. This figure seems excessive if we consider the area conducive to its development. In fact, the data we obtained during our field campaigns plus the calculations made by GIS, show an area of just 1,247 m² (Beato et al., 2017).

The most important patches of vegetation appear towards the S and W, behind the first vegetation band of the marsh. They correspond again with *Zygophyllo fontanesii – Arthrocnemetum macrostachyi* and *Frankenio capitatae – Suadetum verae* that develop on the depressed area of the *malpaís* between 0 and 1 m altitude. The area is covered by clays and sands and has a great salt content. In addition, the greater energy of the water in the southern end of the channel, washes over the *callaos* that border it and creates a level surface of sand that links to the dune space located to the east. Anthropic alteration is intense in this area, which explains why the fixation of the small dunes is produced by substitution thickets (*Chenoleoideo tomentosae – Suaedetum mollis*), which coexist with some halophytic plants and other ruderal fruticosa species. The inclusion of the salt marsh in the urban area of Corralejo is also reflected in the laying down of a mantle of decorative scories and pyroclasts, in which the aforementioned communities are present, as are species of the *Salsolo kali – Cakiletum maritimae and Polycarpo – Nicotianetum glaucae* associations. Finally, different nitrophilous substitution scrubs spread increasingly as we move away from the core of the salt marsh towards the urban perimeter.

In short, the vegetation of the Charco de Britsol salt marsh presents zoning typical of these wetlands and directly related to the level of tolerance the plants have to flooding. Thus, from the foreshore to the supra-steroidal backshore we find: *Halimiono portulacoidis – Salicornietum perennis*, *Zygophyllo fontanesii – Arthrocnemetum macrostachyi*, *Zygophyllo fontanesii – Arthrocnemetum macrostachyi* with *Suaeda vera* and *Frankenio capitatae – Suadetum verae*.

A second distribution structure can be established based on the influence of the sea spray and the plants tolerance of salts. In this case, the coastal rock halophyte shrub (*Frankenio ericifoliae – Zygophylletum fontanesii*) comes first, and behind it, in a second band of vegetation, we find almost monospecific spots of *Suaeda vera*, in addition to nitrophilous sand or arid thickets, depending on the substrate and the level of anthropization (Figure 7).

In addition, we have identified taxa typical of halophilous shrubs (*Atriplex glauca*), of nitro-halophilous sandy grasslands (*Cakile maritima, Salsola kali*), and of nitro-psammophilous scrub (*Polycarpaea nivea*), as well as others of wide distribution such as *Patellifolia patellaris*. Outside the Bristol salt marsh there are also plants typical of urban environments and of nitrified soils, such as *Chenopodium murale*. In total, 31 species have been catalogued, belonging to 25 genera and 12 families. The following taxa are especially noteworthy: *Aizoon canariense*, *Atriplex glauca* var. *ifniensis*, *Atriplex lindeyi*, *Atriplex semilunaris*, *Cuscuta approximata*, *Filago desertorum*, *Lotus glinoides*, *Mesembryanthemum crystallinum*, *Mesembryanthemum nodiflorum*, *Plantago coronopus*, *Senecio leucanthemifolius* and *Sonchus oleraceus* (Beato et al., 2017).

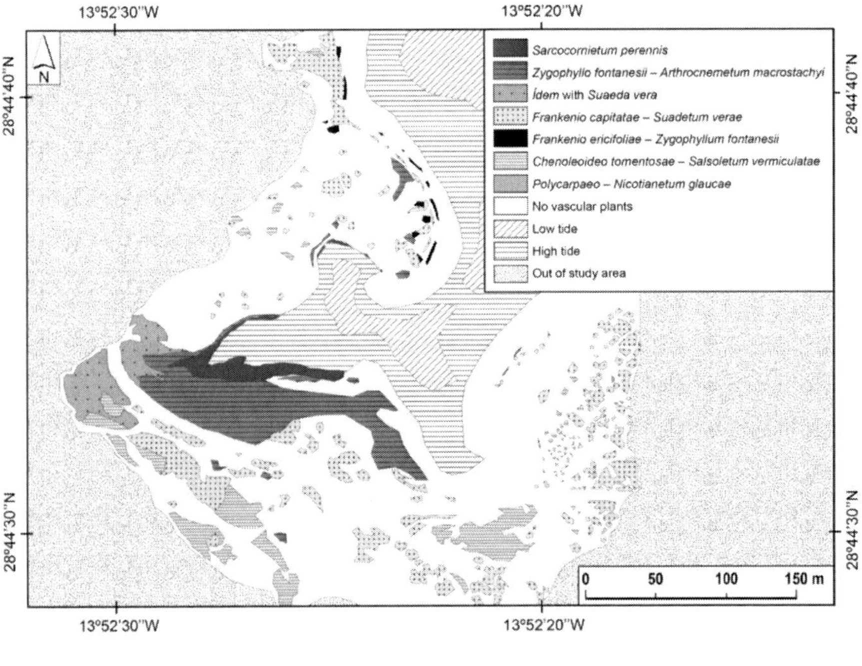

Figure 7. Map of the Charco de Bristol salt marsh (modified from Beato et al., 2017).

The Southern Salt Marshes

The southern end of Fuerteventura, formed by the peninsula of Jandía, is linked to the rest of the island by the Istmo de la Pared. This narrow, elongated isthmus is just 6 km in width, and forms a ridge due to the accumulation of Pliocene-age, wind-blown sands (Criado, 2005).

From the morphostructural point of view, the Jandía peninsula is an old Miocene basaltic massif, with no record of subsequent volcanic activity. It adopts a semicircular structure open to the NNW, known as the Cofete arch, the origin of which is associated to large debris-avalanches that the old Jandía volcano experienced (20.7-14.2 Ma). This volcano was between 12 and 15 km wide and reached up to 2,000 m a.s.l. (Cubas et al., 1992). To the south, the excavation of deep ravines has generated a set of crest-shaped interfluves known locally as *cuchillos* (knives).

The largest salt marshes of the Macaronesian region can be found in this area. They are situated in the Sotavento and the Matorral beaches, within the municipality of Pájara, and are habitats of great faunal importance due, in particular, to the presence of migratory birds and endemisms that include, among others, *Actitis hypoleucos, Arenaria interpres, Calandrella rufescens, Calidris alba, Calidris alpina, Charadrius hiaticula, Larus marinus, Numenius* (*arquata and phaeopus*), *Pluvialis squatarola, Saxicola dacotiae dacotiae, Stercorarius parasiticus* and *Tringa* (*erythropus, nebularia, totanus*). The marine areas of these enclaves are protected by the Playa de Sotavento de Jandía SAC, both because of the presence of nesting *Caretta caretta* and *Chelonia mydas* sea turtles, and because of the presence of the dolphin *Tursiops truncatus*. Sadly, until recently these areas have been vulnerable to and severly damaged by the tourists boom of the last few decades.

The Playa de Sotavento Salt Marsh

The Playa de Sotavento, also known as Saladar de Jandía, is located on the eastern slope of the La Pared isthmus. It is a set of beaches or sandy areas separated by small cliffs, some 15 km long and more than 500 m wide. Along its outer edges, it is affected by various roads and tourist developments, and marine aggregates extraction has altered the natural dynamics of the enclave. However, its inclusion in the Jandía SAC, and its closure to private vehicles explains its current healthy state, which only put at risk by the huge influx of visitors, and by possible sea spills if certain oil-industry initiatives prosper.

From the geomorphological point of view, the most outstanding feature of this beach is the sand bar that extends parallel to the shoreline for almost 4 km. This sand bar has several mouths or inlets that facilitate the ebb and flow of seawater with the tide, resulting in a coastal lagoon being formed between the cliff and the sand bar (Alonso et al., 2002). Made up of muddy sand, it is flooded during high tide and exposed during the low spring tide. A halophilous habitat develops in the proximity of the lagoon, and even inside it, as a result of which, *Arthrocnemum macrostachyum* bushes appear partially moistened and completely surrounded by sea water at high tide. Small elevations that allow the stabilization of other stands of vegetation isolated by water also develop.

The vegetation of this salt marsh is dominated by *Zygophyllo fontanesii – Arthrocnemetum macrostachyi*, which is distributed around Casas del Risco del Paso in three large masses to the south (2, 5 and 6 ha, respectively), and two to the north (2 and 7 ha), as well as one in the vicinity of the urbanization Los Verodes (21 ha). Inland, when on sands, the salt marsh is enriched with *Traganum moquinii*, and by halo-psammophile and nitro- psammophile thicket species of *Polycarpaeo niveae – Lotetum lancerottensis* and *Salsola divaricata*. The more anthropized areas are evidenced by the presence of taxa of the *Chenoleoideo tomentosae – Suaedetum mollis*.

The Playa del Matorral Salt Marsh

The Playa del Matorral, located on the southern tip of Fuerteventura near the town of Morro Jable, is formed by marine origin, organogenic sands, and fine volcanic sediments mainly contributed by the Vinamar gorge. Geomorphologically, it is a small delta formed after a long process of evolution; initially, the confluence of sand bars that originated a coastal lagoon, and subsequently, the filling of the lagoon by sediments from the discharge of the three ravines that flow into it. At present, it has an area of around 100 ha, and is basically flat, but presents slight irregularities which allow the formation of channels and pools after the entry of sea water during high tide. Thus, the vegetation that colonizes this delta-shaped wetland is made up of halophilic communities and species resistant to recurrent flooding. It shows fairly dense coverage, although this is fragmented by numerous paths and tracks, and old debris storage areas. In addition, the areas most affected by the marine floods are devoid of vegetation.

The dominant plant community is *Zygophyllo fontanesii – Arthrocnemetum macrostachyi*, which occupies the backshore and upper limit of the foreshore, whilst the shortage of *Tetraena fontanesii* in all the inventories and transects is of note. Also of note is the abundance of *Suaeda vera* which, together with *Arthrocnemum macrostachyum*, creates dense carpets in frank expansion.

To the northwest of the lighthouse, in the central section of the Playa del Matorral, *Arthrocnemum macrostachyum* constitutes a practically monospecific formation, both in isolated spots and in continuous coverage, and forms an even carpet only broken by the presence of various specimens of the taller *Tamarix canariensis*. To the west, *Frankenia capitata* is scattered over open spaces devoid of vegetation, together with certain specimens of *Tetraena fontanesii*, and with shrubs of *Arthrocnemum macrostachyum* and other species tolerant of sporadic flooding and high salinity (Figure 8).

Figure 8. The Playa del Matorral salt marsh at Morro Jable (Pájara).

To the east of the lighthouse, *Suaeda vera* occupies large areas above the backshore, and therefore sheltered from the daily tides. It spreads easily, colonizing new spaces and threatening the dominance of *Arthrocnemum macrostachyum* in certain areas where the *Frankenio capitatae – Suaedetum verae* community prevails. However, it forms a more or less continuous vegetal cover in which both species alternate, both on sands and on clay and silty bottoms, leaving little space for other taxa.

The large size of the salt marsh plus the existence of several biotopes with a range of edaphic characteristics, explains the richness of the plant heritage. Thus, we can find numerous species, such as *Aizoon canariense*, *Atriplex halimus*, *Desmanthus virgatus*, *Phoenix canariensis*, *Rumex vesicarius*, *Salsola divaricata*, *Sclerocephalus arabicus*, *Suaeda maritima*, and *Tamarix canariensis*, as well as *Policarpea nivea* and *Traganum moquinii* in sandy areas away from the direct influence of the sea.

In addition, due to the severe environmental deterioration that the area has suffered, different forms of regression scrub can be distinguished, such as the association *Chenoleoideo tomentosae – Suaedetum molli* (found over a wide area and with a strong presence of *Launaea arborescens*) and *Nicotiana glauca*. These occupy large spaces in the urban limits of the salt marsh, and are linked to anthropic elements, particularly surrounding walkways, tracks, buildings, remains of tailings from the central sector of the marsh, as well as to the mouth of the Vinamar gorge. The high number of allochthonous species, and even invasive species from gardening, needs mentioning. These include *Acacia cyanophylla*, *Argyranthemum frutescens frutescens*, *Carpobrotus edulis*, *Casuarina equisetifolia*, *Kikuyuochloa clandestina*, *Limoniastrum monopetalum*, *Pennisetum clandestinum, Phragmites australis* and *Sesuvium portulacastrum*.

Other Salt Marshes of Fuerteventura

There are formations of *Zygophyllo fontanesii – Arthrocnemetum macrostachyi* distributed discontinuously along the coastline on both the western and eastern slopes of the island. On the beaches to the south of El

Cotillo, the salt marsh is sheltered in the crevices and small gullies that dissect the cliff. There are no buildings in the area, but it is much frequented by tourists and it receives significant traffic from private vehicles.

On the Playa del Castillo, the salt marsh occupies a gully that is strongly anthropized by different runoffs. About 300 m long and with an area of 4 ha including its different branches, it has steep sides and an almost flat bottom (Figure 9). The taxa of the association *Chenoleoideo tomentosae – Suaedetum mollis* not only abound in the escarpments but also in the center of the marsh, alongside other species of *Polycarpo – Nicotianetum glaucae* on debris deposits and land that has been turned over. *Arthrocnemum macrostachyum* dominates in the most humid and brackish areas, while *Suaeda vera* does so along the outer edge. At the mouth of these incipient ravines, sand from the beach favours the development of psammophiles of the associations *Polycarpaeo niveae – Lotetum lancerottensis*, *Traganetum moquini* and *Euphorbio paraliae – Cyperetum capitati*, with the scattered presence of *Tetraena fontanesii*. This distribution is repeated in the smaller badlands located to the south, although substitution scrub is less important, and is limited to the borders of the area. In general, they always have a monospecific nucleus of *Arthrocnemum macrostachyum* guarded by equally dense but smaller-sized spots of *Suaeda vera*.

Figure 9. Salt marsh vegetation and substitution scrubs in a gully, heavily anthropized, on the cliff of the Playa del Castillo beach (La Oliva).

The Aljibe de la Cueva beach is home to a huge gully some 2 ha in size. With a sub-horizontal bed excavated into the cliff, the gully is colonized by salt marsh vegetation. It is characterized by the monospecific dominance of *Arthrocnemum macrostachyum,* which covers the depression as a carpet up to 1.5 meters high. The edges of the gully are covered by abundant *Suaeda vera* thickets and small regressive series shrubs. In addition, there are hygrophilous communities of *Enteromorpho intestinalis – Ruppietum maritimae* in small ponds at the mouth of the ravine, as well as the psammophiles mentioned above in the areas covered by sand, and samples of *Astidamia latifolia.* This situation is also found in another large gully to south, which has a sandy inlet where *Arthrocnemum macrostachyum, Traganum moquinii* and *Suaeda vera* immobilize small dunes. Inland, monospecific spots of *Arthrocnemum macrostachyum* surrounded by *Suaeda vera* cover various branches of the gully (Figure 10).

Figure 10. Salt marsh vegetation in a gully on the cliff of the Aljibe de la Cueva beach (La Oliva).

In Punta Toneles, in the east of Fuerteventura, *Zygophyllo fontanesii – Arthrocnemetum macrostachyi* is distributed in scattered groups over a basaltic lava flow that descends over an inactive cliff and runs into the sea. The horizontal nature and low altitude of this recent coastal area has favored the deposition of clays and sands both in any small depressions, and in the mouth of the La Cueva gorge. In these places *Arthrocnemum macrostachyum* finds suitable conditions, and benefits from the difficult access and the almost zero anthropization.

Finally, *Frankenio capitatae – Suadetum verae* is also widely represented on the coast of Fuerteventura. Its protagonism has already been mentioned in various salt marshes in the north of the island, but its location in gorges where brackish waters flow also needs mentioning. This is the case with the Los Molinos gorge, where it extends for some 4 km, and travelling inland more than 3 km from the coast. It is found in other minor gullies on both the west and east coasts, mainly in gully mouths influenced by the sea. Lastly, it can also be found inland, as in the Muley gorge, where *Suaeda vera* runs along the bed of the gorge for more than 3 km in dense, isolated strips.

Management and Threats in the Salt Marshes of Fuerteventura

The factor responsible for the degradation of the coastal areas of the Canary Islands has been the model of economic development based on mass tourism and the resulting uncontrolled urbanization (Fernández-Cabrera et al., 2011). This has altered the functions and the ecological processes of the salt marshes, and has finally produced a loss of both the natural and cultural heritage (Pérez-Chacón et al., 2007). Practically all the recent territorial changes in Fuerteventura, including indirectly the salt marshes, have their origin in the tourist boom and its repercussions on the dynamics of the island's demographics and economy (García-Romero et al., 2016; Beato et al., 2017). According to the data provided by the Canary Islands Institute of Statistics, tourism on Fuerteventura has experienced a huge progression in recent decades, reaching more than three million visitors in 2016 (far exceeding the record of 1 million tourists from 2011), with Corralejo and the beaches of the south and of Sotavento as the main destinations. In terms of population, Fuerteventura has gone from 37,000 people in 1991 to over 107,000 in 2016, an increase of 289% in 25 years.

Coastlines are hugely fragile areas, a property that is even more acutely true of small, coastal wetlands located within the area of influence of tourism resorts. As a result of this, the current situation of the Fuerteventura salt marshes heavily depends on their spatial relationship with the urbanized areas in their surroundings, as with Morro Jable and Corralejo, regardless of any protection they may receive to a greater or lesser extent from the existing figures of conservation. Thus, in the south of the island, the Playa del Matorral salt marsh has suffered serious degradation due to urban growth around Morro Jable, whilst the salt flats of the Sotavento beach are well preserved thanks to limited building. In contrast, the Charco de Bristol marsh warrants special attention; not protected in any way, it is home to some unique island heritage, but has suffered serious deterioration due to the urban development of Corralejo (Figure 11).

The salt marsh of the Playa del Matorral was named the El Saladar Natural Landscape of National Interest on June 19, 1987, and reclassified as a SSI on December 19, 1994. It became a Macaronesian biogeographic region protected area within the European Natura 2000 Network, when it was designated as a SPA on October 17, 2006, and as a SAC on December 29, 2010. In addition, it is the only Canarian salt marsh to have been part of the Ramsar List since 2002. However, the marsh is a heavily anthropized space as a result of the exploitation of sea salt between 1935 and 1950, from when a small lifting tower remains. Later, in the 70s, the first tourism facilities were set up, notably the Stella Canaris resort. This would be followed by a multitude of actions that would completely degrade the landscape.

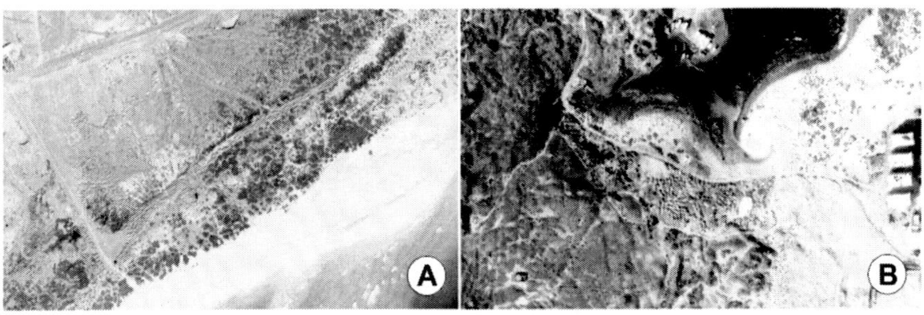

Figure 11. Aerial photographs of La Playa del Matorral (A) and Charco de Bristol (B) of the 1973-1986 Interministerial Flight taken in January 1981. Large areas can be seen occupied by mounds of debris deposits from the construction industry.

In recent years, a LIFE project for ecological recovery has brought about the elimination of 90,000 m^3 of debris left by the construction of tourism infrastructures and accommodation, together with walls and other stone and concrete constructions from the extraction of sea salt mentioned above. Garbage, waste, and invasive species have also been eliminated, and 200 palm trees extracted. These and other efforts aimed at reducing the impact of the large influx of tourists to the area have improved the environment of the marsh.

A management plan has been in place since 2016, and sets out four different zones: priority conservation, conservation, restoration and transition. The 50.6 ha of the priority conservation zone aims to protect and conserve Habitat 1420, Mediterranean and thermo-Atlantic halophilous shrubs (*Sarcocornetea fruticosae*). It includes the flood zone between the salt marsh and the beach to the SW of the SAC. The conservation zone also contains areas of great value due to the presence of species or habitats of Community interest, although their present poor state of conservation requires certain restoration measures. The zone presents scant coverage of *Sarcocornetea fruticosae*, has an area of 1.4 ha and corresponds to the NW sector of the Playa del Matorral marsh. The 8.2 ha restoration zone includes areas made up of degraded or substitution facies that could be recovered. This zone is covered by the phytosociological associations *Chenoleoideo tomentosae – Suadetum mollis, Suaedo verae – Tamaricetum canariensis* and by landscaped areas. Finally, the transition zone, with an area of 35.3 ha, supports uses and activities that have caused a high degree of environmental change, and consists of the paths and access roads to the beach, the road and the perimeter walk, and the tourism facilities that delimit the north and west ends of the salt marsh (Figure 12).

The conservation measures adopted for this whole marsh include the prohibition of the dumping of rubble and building material, the introduction of non-native species of flora and fauna, surface dumping and the movement and extraction of soil, plus any activity that increases erosion processes in the area of influence of the tide or at the mouth of gorge. Only scientific, information, and educational activities are allowed, provided they are carried out along the paths that cross the area, and do not involve the use of motor vehicles. Noteable among the actions carried out recently, are: the updating of the cartography of the habitats of Community interest and any remaining units of vegetation, the assessment of the state of conservation, the erection of exclusion fencing around the different habitats of the salt marsh, signposting and information panels explaining the natural resources, the cleaning and removal of rubbish, the eradication of invasive species, and the elimination of effluents and waste (Carrillo, 2016).

Nonetheless, Playa del Matorral salt marsh is still subject to multiple threats, such invasive alien species of flora and fauna: the former include giant cane (*Arundo donax*), hottentot-fig (*Carpobrotus edulis*), and the Kikuyuochloa clandestina grass in the vicinity of the landscaped areas (Martín et al., 2009); the latter include the barbary ground squirrel (*Atlantoxerus getulus*) and the monk parakeet (*Myiopsitia monachus*), attracted by the remains of food that tourists leave. Other threats come from transit and trampling by tourists using the salt marsh for sunbathing, despite the fencing and the prohibition signs.

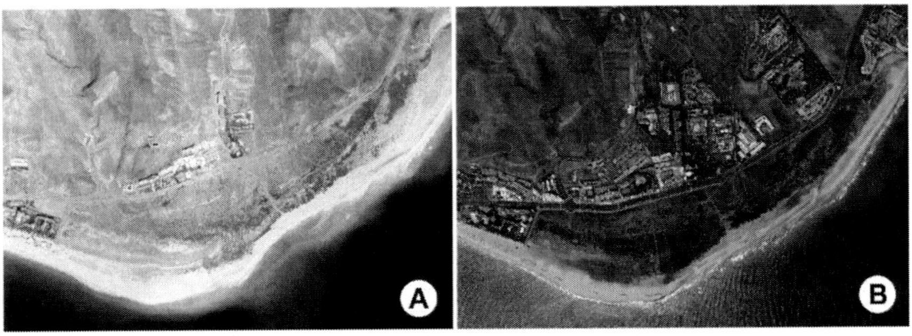

Figure 12. Aerial photographs of La Playa del Matorral. (A) 1973-1986 Interministerial Flight, taken in January 1981. (B) 2015 orthophoto map of the National Plan of Aerial Orthophotography. Note the expansion of buildings constricting the protected space.

The other salt marsh of great scientific interest on the island, the Charco de Bristol, has been affected by the demographic growth and tourist expansion of Corralejo, which has gone from a built area of 6.72 ha in 1969 to almost 321 ha in 2009 (Fernández-Cabrera et al., 2011). This new building has constricted the whole of the Dunas de Corralejo e Isla de Lobos Natural Park and has reached its boundaries. In the aerial photographs of 1981 (Institute of Agrarian Reform and Development) the first blocks of apartments can be seen in the vicinity of the marsh, on what today is the final section of the Avenida Juan Carlos I. Also visible is the use of the entire southern end of the marsh as a dump or deposit for a wide range of materials. After these activities ended, the remaining materials were spread over the original deposits, that is, the organogenic sands and clays, radically disturbing the plant carpet. At present, the most disruptive element of the Charco de Bristol salt marsh is the track that crosses it. Already present in 1945 as a local road, it borders the north of the island from the Faro del Toston to Corralejo. It supports relatively dense traffic and crosses one of the most significant plant communities of the Charco de Bristol, the *Zygophyllo fontanesii – Arthrocnemetum macrostachyi*. It also facilitates road access to all the plant communities of the marsh.

Although it is true that the municipality of La Oliva has been aware of the ecological value of the Charco salt marsh in Bristol in recent years, it has not yet put into practice any effective measure aimed at its recovery. Ten years ago, for example, they were behind the opening of another road for traffic outside the marsh (which turned out to be illegal on having been built on private property), the installation of an information panel (showing basically the name of the enclave), a stake fence that was never completed, as well as the laying down of a mantle of scories and pyroclasts on the eastern beach (that introduced foreign plant species). On the other hand, they have also collaborated in cleaning actions initiated occasionally by Corralejo's own neighborhood groups, but which are clearly insufficient for the conservation of the marsh.

Recently Beato et al. (2017) formulated a management proposal for the Charco de Bristol salt marsh based, first, on its declaration as a SSI, and then on the application of restoration measures. The proposal sets out five zones: an exclusion zone, restoration zone 1, restoration zone 2, restoration zone 3, and a moderate use zone. The exclusion zone corresponds to the central part of the salt marsh where the population of *Sarcocornia perennis* and the main patch of *Zygophyllo fontanesii – Arthrocnemetum macrostachyi* are found. The aim of the zone is to eliminate the indiscriminate movement of people and domestic animals along present paths. The reduction in trampling and soil compaction would allow the gradual recovery of the zone, and its connection with different vegetal formations. Restoration zone 1 corresponds to the area colonized by the association *Zygophyllo fontanesii – Arthrocnemetum macrostachyi* in coexistence with *Suaeda vera*. It is separated from the centre of the marsh by a track, and houses a rubbish dump. The elimination of the dump is proposed first of all, and once restored it would become part of the exclusion zone. Restoration zone 2 coincides with the track that crosses the salt marsh. The first measure is the suppression of road traffic and the creation of a wooden walkway with lookouts and information panels, as well as the regeneration of the ground the track is on. Once completely restored, it would become part of the zone of moderate use. Restoration zone 3 houses various plant species of great interest such as *Arthrocnemum macrostachyum* and *Limonium papillatum*. It has been affected by the entry of vehicles and the dumping of construction debris. The recommendation here is the prohibition of traffic, however slight, and the elimination of urban waste and the remains of any buildings (blocks and concrete posts, parts of metal structures, etc.). Finally, the zone of moderate use, corresponding to the eastern outer edge of the salt marsh, would be given over to recreational and pedagogical purposes.

The threats that loom over the Charco de Bristol salt marsh are still numerous and jeopardize its future. Fuel spills, dumping and the accumulation of construction debris, indiscriminate pedestrian and vehicle traffic, garbage and pet waste, to which we now have to add the Corralejo City Council project to build natural outdoor swimming pools.

CONCLUSION

Salt marshes are very rare habitats in the Canary Islands archipelago and, in general, in the Macaronesian area. However, Fuerteventura hosts truly significant wetlands from the scientific point of view. In fact, the salt marshes in Fuerteventura currently occupy an area of 132 ha, which is almost half the surface of all of the Canary Islands salt marshes. This type of habitat and its plant formations are heavily fragmented and scattered along the coast, and so display a wide variety from the geomorphological point of view. More specifically, we have distinguished seven types of salt flats in Fuerteventura: open coast, lagoon, delta, restricted-entrance embayment, low coast of aa lava flows, low cliff with gullies and lower courses, and the mouths of gorges. These salt marshes are characterized by their high ecological quality and great floristic richness, with xerophile, psammophile, hygrophile and halophile plants integrated in *Sarcocornietea fruticosae*. They belong to the thermo-Atlantic halophilic ecosystem and have been catalogued as code 1420 habitats of European Community interest. The main plant formations in these brackish environments are *Zygophyllo fontanesii – Arthrocnemetum macrostachyi* and *Frankenio capitatae – Suaedetum verae*, which occupy the upper limit of the foreshore and backshore, respectively, and the abundance of regressive series plants such as *Chenoleoideo tomentosae – Suaedetum molli* and *Chenoleoideo tomentosae – Salsoletum vermiculatae* resulting from the intense anthropic degradation the marshes have experienced. In contrast, in the Charco de Bristol the main feature is the presence of *Sarcocornia perennis*, which occupies the lower levels of the salt marsh, that is, the transition zone between the foreshore and the backshore, and which is flooded twice daily during high tide. This population of *Sarcocornia perennis* belonging to the community *Halimiono portulacoidis – Salicornietum perennis* is unique in the whole of Fuerteventura, and is included in the Canary Islands' Catalogue of Threatened Species as being of high interest.

In general, the salt marshes of Fuerteventura are threatened by the constant pressure exherted on them by the apparently unstoppable urban and tourism growth that the island has experienced over the last fifty years. In particular, the urban-tourist complexes of Corralejo and Morro Jable threaten the ecological value and floristic richness of two salt marshes of maximum scientific interest, the Charco de Bristol and the Playa del Matorral marshes, respectively. The Playa del Matorral salt marsh, protected as a SSI since 1994, is in a clear process of recovery, and its survival is assured. In contrast, the Charco de Bristol salt marsh lacks any form of protection whatsoever, and is in a precarious state of conservation due to the multiple aggressions that it suffers daily. Because of this, its declaration as a SSI, and the application of urgent restoration measures to prevent its permanent disappearance, is of paramount importance.

REFERENCES

Adnitt, C., Brew, D., Cottle, R., Hardwick, M., John, S., Legget, D., McNulty, S., Meakins, M., & Staniland, R. (2007). *Saltmarsh management manual*. Bristol: Environment Agency.

Allen, J. R. L. (2000). Morphodynamics of Holocene salt marshes: A review sketch from the Atlantic and Southern North Sea coasts of Europe. *Quaternary Science Reviews* 19, 1155-1231.

Alonso, I., Alcántara-Carrió, J., & Cabrera, L. (2002). Tourist Resorts and their Impact on Beach Erosion at Sotavento Beaches, Fuerteventura, Spain. *Journal of Coastal Research* SI 36, 1-7.

Ancochea, E., Brandle, J. L., Cubas, C. R., Hernán, F., & Huertas, M. J. (1993). *La Serie I de la isla de Fuerteventura*. Madrid: Memoria de la Real Academia de Ciencias Exactas, Físicas y Naturales. [*The old basaltic series of the island of Fuerteventura*. Madrid: Memory of the Royal Academy of Exact, Physical and Natural Sciences].

Arozena, M. E. (2000). Estructura de la vegetación. In G. Meaza (Ed.), *Metodología y práctica de la Biogeografía* (pp. 77-146). Madrid: Ediciones del Serbal. [Structure of vegetation. In G. Meaza (Ed.), *Methodology and practice of Biogeography* (pp. 77-146)].

Beato Bergua, S., Poblete Piedrabuena, M. A., & Marino Alfonso, J. L. (2017). El saladar de Bristol: patrimonio vegetal, estado de conservación y propuestas de restauración (Corralejo, Fuerteventura, Islas Canarias). *Boletín de la Asociación de Geógrafos Españoles* 73, 223-246. [Bristol salt marsh: vegetal heritage, conservation status and restoration proposal (Corralejo, Fuerteventura, Canary Islands). *Annals of the Association of Spanish Geographers* 73, 223-246].

Beltrán, E., Arozena, M. E., & Ríos, J. (1999). El papel de la morfología superficial de las coladas lávicas en las características de la vegetación del malpaís de Güímar (Tenerife, I. Canarias). In *Professor Joan Vilà Valentí* (pp. 783-799). Barcelona: Universitat de Barcelona. [The role of the surface morphology of the lava flows in the vegetation characteristics of the malpaís de Güímar (Tenerife, I. Canarias). In *Professor Joan Vilà Valentí* (pp. 783-799). Barcelona: University of Barcelona].

Braun-Blanquet, J. (1979). *Fitosociología. Bases para el estudio de las comunidades vegetales*. Madrid: Blume. [*Phytosociology. Bases for the study of plant communities*].

Carrillo, J. (2016). *Plan de Gestión de la Zona Especial de Conservación ES7010042 La Playa del Matorral (Fuerteventura).* Fuerteventura: Consejería de Política Territorial, Sostenibilidad y Seguridad. Gobierno de Canarias. [*Management Plan of the Special Conservation Area ES7010042 The Matorral Beach (Fuerteventura).* Fuerteventura: Ministry of Territorial Policy, Sustainability and Security. Government of the Canary Islands].

Coello, J., Cantagrel, J. M., Hernán, F., Fúster, J. M., Ibarrola, E., Ancochea, E., Casquet, C., Jamond, C., Díaz de Terán, J. R., & Cendrero, A. (1992). Evolution of the Eastern Volcanic Ridge of the Canary Islands Based on New K-Ar Data. *Journal of Volcanology and Geothermal Research* 53, 251-274.

Criado, C. (1990). La evolución del paisaje vegetal de Fuerteventura a través de fuentes escritas (s. XV – XIX). *Tebeto (Anuario del Archivo Histórico Insular de Fuerteventura, Islas Canarias)* 3, 249-259. [The evolution of the vegetal landscape of Fuerteventura through written sources (XV - XIX centuries). *Tebeto (Yearbook of the Insular Historical Archive of Fuerteventura, Canary Islands)* 3, 249-259.]

Criado, C. (1991). *La evolución del relieve de Fuerteventura*. Puerto del Rosario: Servicio de Publicaciones del Cabildo Insular de Fuerteventura. [*The evolution of the relief of Fuerteventura*].

Criado, C., Guillou, H., Hansen, A., Hansen, C., Lillo, P., Torres, J. M., & Naranjo, A. (2004). Geomorphological evolution of Parque Natural de

Las Dunas de Corralejo (Fuerteventura, Canary Islands). In G. Benito & A. Díez-Herrero (Eds.), *Contribuciones Recientes sobre Geomorfología* (pp. 291-297). Madrid: Sociedad Española de Geomorfología y Consejo Superior de Investigaciones Científicas. [In G. Benito & A. Díez-Herrero (Eds.), *Recent Contributions on Geomorphology* (pp. 291-297). Madrid: Spanish Society of Geomorphology and Higher Council of Scientific Research].

Criado, C. (2005). Formas de modelado y procesos morfogenéticos. In O. Rodríguez (Coord.), *Patrimonio Natural de la isla de Fuerteventura* (pp. 45-58). Tenerife: Cabildo de Fuerteventura. Centro de la Cultura Popular Canaria. Consejería de Medio Ambiente y Ordenación Territorial. [Landforms and morphogenetic processes. In O. Rodríguez (Coord.), *Natural Heritage of the island of Fuerteventura* (pp. 45-58). Tenerife: The Island Council of Fuerteventura. Center for Canarian Popular Culture. Department of Environment and Territorial Planning].

Criado, C., Guillou, H., Hansen, A., Hansen, C., Lillo, P., Torres, J. M., & Naranjo, A. (2007). Mapa Geomorfológico del Parque Natural de las Dunas de Corralejo (Fuerteventura). *Actas de las XIII Jornadas de estudio de Lanzarote y Fuerteventura* (399-427). Puerto del Rosario: Servicio de Publicaciones Cabildo de Fuerteventura – Cabildo de Lanzarote. [Geomorphological Map of the Natural Park of the Dunes of Corralejo (Fuerteventura). *Proceedings of the XIII Study Conferences of Lanzarote and Fuerteventura* (399-427).].

Criado, C., Torres, J. M., Hansen, A., Lillo, P., & Naranjo, A. (2012). Intercalaciones de polvo sahariano en paleodunas bioclásticas de Fuerteventura (Islas Canarias). *Cuaternario y Geomorfología* 26 (1-2), 73-88. [Intercalations of Saharan dust in bioclastic paleodunes of Fuerteventura (Canary Islands). *Quaternary and Geomorphology* 26 (1-2), 73-88.

Cubas, C. R., Fernández, S., Hernández-Pacheco, A., Hernán, F., & de la Nuez, J. (1988). Los domos sálicos de Fuerteventura. *Revista de materiales y procesos geológicos* 6, 71-97. [The salic domes of Fuerteventura. *Journal of materials and geological processes* 6, 71-97].

Cubas, C. R., Hernán, F., & Ancochea, E. (1992). El edificio sur (Jandía) de la serie I de Fuerteventura: rasgos generales. *Geogaceta* 11, 79-81. [The southern (Jandia) edifice in the old basaltic series of Fuerteventura: general features. *Geogazette* 11, 79-81].

Curtis, T. G. F., & Sheehy Skeffington, M. J. (2008). The salt marshes of Ireland: an inventory and account of their geographical variation. *Biology and Environment: Proceedings of the Royal Irish Academy* 98B: 87-104.

Del Arco, M. J., Wildpret, W., Pérez de Paz, P. L., Rodríguez, O., Acebes, J. R., García, A., Martín, V. E., Reyes, J. A., Salas, M., Bermejo, J. A., González, R., Cabrera, M. V., & García, S. (2006). *Mapa de Vegetación de Canarias*. Santa Cruz de Tenerife: GRAFCAN. [*Vegetation map of the Canary Islands*].

Del Arco, M. J., González-González, R., Garzón-Machado, V., & Pizarro-Hernández, B. (2010). Actual and potential natural vegetation on the Canary Islands and its conservation status. *Biodiversity and Conservation* 19, 3089-3140. https://doi.org./10.1007/s10531-010-9881-2.

Del Arco, M. J., Izco, J., & Rodríguez, O. (2017). The name *Salicornietum perennis* (*sensu lato*) and global nomenclatural issues. *Lazaroa* 38(2), 203-206. http://dx.doi.org/10.5209/LAZA.56609.

Díaz, G. (2007). *Ficha Informativa Ramsar. Saladar Jandía*. Las Palmas de Gran Canaria: Gobierno de Canarias. [*Ramsar Information Sheet. Saladar Jandía*. Las Palmas de Gran Canaria: Government of the Canary Islands]. Retrieved from the Government of the Canary Islands website: http://www.gobiernodecanarias.org/opencmsweb/export/sites/medioambiente/piac/galerias/descargas/Documentos/Biodiversidad/Biodiversidad_ficha_ramsar.pdf.

Dijkema, K. S. (1987). Geography of salt marshes in Europe. *Zeitschrift für Geomorphologie* 31, 489-499. [*Journal of Geomorphology*]

Dóniz, F. J. (2009). Patrimonio geomorfológico de los volcanes basálticos monogénicos de la caldera de Gairía-Malpaís Chico y el Malpaís Grande en la isla de Fuerteventura (Canarias, España). *Nimbus* 23-24, 89-103. [Geomorphological heritage of the monogenic basaltic volcanoes of the caldera of Gairía-Malpaís Chico and the Malpaís Grande on the island of Fuerteventura (Canary Islands, Spain). *Nimbus* 23-24, 89-103].

Doody, J. P. (2008). *Saltmarsh Conservation, Management and Restoration. Coastal Systems and Continental Margins, 12.* Springer.

Dorta, P. J. (1996). Las inversiones térmicas en Canarias. *Investigaciones Geográficas* 15, 109-126. [Temperature inversions in the Canary Islands. *Geographic Investigations* 15, 109-126].

Espinar, J. L. (2009). *1420 Matorrales halófilos mediterráneos y termoatlánticos (Sarcocornetea fruticosi).* Madrid: Ministerio de Medio Ambiente, y Medio Rural y Marino. [*1420 Mediterranean halophilic scrub and thermoatlantic scrub (Sarcocornetea fruticosi).* Madrid: Ministry of the Environment, and Rural and Marine Affairs].

Fernández, E., Roca, E., Cabrera, L., Hernández-Calvento, L., & Pérez-Chacón, E. (2012). Estudio de la percepción social en el entorno del Parque Natural de las Dunas de Corralejo (Fuerteventura, Islas Canarias): aplicaciones para la gestión integrada de zonas costeras. In J. A. Chica, J. García, J. García, M. L. Pérez (Eds.), *Mejorando la gestión de las áras litorales de Iberoamérica* (pp. 1186-1196). Cádiz: Grupo de Investigación Gestión Integrada de Áreas Litorales. [Study of the social perception in the environment of the Natural Park of the Dunes of Corralejo (Fuerteventura, Canary Islands): applications for the integrated management of coastal zones. In J. A. Chica, J. García, J. García, M. L. Pérez (Eds.), *Improving the management of the littoral areas of Ibero-America* (pp. 1186-1196). Cádiz: Research Group Integrated Management of Coastal Areas.

Fernández, M., & Santos, A. (1983). La vegetación del litoral de Canarias, I. Arthrocnemetea. *Lazaroa* 5: 143-155. [*The vegetation of the Canary coast, I. Arthrocnemetea. Lazaroa* 5, 143-155].

Fernández-Cabrera, E., Pérez-Chacón, E., Cruz, N., Hernández-Cordero, A., & Hernández-Calvento, L. (2011). Consecuencias ambientales del crecimiento urbano-turístico en el sistema de dunas de Corralejo (Fuerteventura-Islas Canarias). In V. González & J. A. Marco (Eds.), *Urbanismo expansivo: de la utopía a la realidad* (pp. 241-252).

Alicante: Universidad de Alicante. [Environmental consequences of urban-tourist growth in the dune system of Corralejo (Fuerteventura-Canary Islands). In V. González & J. A. Marco (Eds.), *Expansive urbanism: from utopia to reality* (pp. 241-252). Alicante: University of Alicante].

Fúster, J. M., Cendrero, A., Gastesi, P., Ibarrola, E., & López-Ruiz, J. (1968). *Geología y vulcanología de las Islas Canarias. Fuerteventura.* Madrid: Instituto Lucas Mallada. [*Geology and volcanology of the Canary Islands. Fuerteventura*]

García-Romero, L., Pérez-Chacón, E., Hernández-Cordero, A. I., & Hernández-Calvento, L. (2014). Cartografía de la distribución espacial de la cobertura vegetal en sistemas arenosos áridos: análisis comparado de métodos aplicados en La Graciosa (Islas Canarias, España). In R. Cámara, B. Rodríguez & J. L. Muriel (Eds.), *Biogeografía de sistemas litorales. Dinámica y conservación* (pp. 21-26). Sevilla: Universidad de Sevilla. [Cartography of the spatial distribution of vegetation cover in arid sandy systems: comparative analysis of methods applied in La Graciosa (Canary Islands, Spain). In R. Cámara, B. Rodríguez & J. L. Muriel (Eds.), *Biogeography of coastal systems. Dynamics and conservation* (pp. 21-26). Sevilla: University of Sevilla.

García-Romero, L., Hernández-Cordero, A. I., Fernández-Cabrera, E., Peña-Alonso, C., Hernández-Calvento, L., & Pérez-Chacón, E. (2016). Urban-touristic impacts on the Aeolian sedimentary systems of the Canary Islands: conflict between development and conservation. *Island Studies Journal* 11 (1), 91-112.

Hernández-Cordero, A. I, Pérez-Chacón, E., & Hernández-Calvento, L. (2014). Dinámica de la comunidad de Tamarix Canariensis en el campo de dunas de Maspalomas (Gran Canaria, Islas Canarias). In R. Cámara, B. Rodríguez & J. L. Muriel (Eds.), *Biogeografía de sistemas litorales. Dinámica y conservación* (pp. 15-19). Sevilla: Universidad de Sevilla. [Dynamics of the community of Tamarix Canariensis in the dune field of Maspalomas (Gran Canaria, Canary Islands). In R. Cámara, B. Rodríguez & J. L. Muriel (Eds.), *Biogeography of coastal systems. Dynamics and conservation* (pp. 15-19). Sevilla: University of Sevilla].

Hernández-Cordero, A. I., Pérez-Chacón, E., & Hernández-Calvento, L. (2015). Vegetation, distance to the coast, and Aeolian geomorphic processes and landforms in a transgressive arid coastal dune system. *Physical Geography* 36 (1), 60-83.

Ibarrola, E., Fúster, J. M., & Cantagrel, J. M. (1989). Edades K-Ar de las rocas volcánicas submarinas en el sector norte del Complejo Basal de Fuerteventura. *Proceedings of ESF Meeting on Canarian Volcanism* (pp. 124-128). Lanzarote: European Science Foundation – Consejo Superior de Investigaciones Científicas. [K-Ar ages of underwater volcanic rocks in the northern sector of the Basal Complex of Fuerteventura. *Proceedings of the ESF Meeting on Canarian Volcanism* (pp. 124-128). Lanzarote: European Science Foundation - Superior Council of Scientific Investigations].

Kunkel, G. (1977). Las plantas vasculares de Fuerteventura (Islas Canarias), con especial interés de las forrajeras. *Naturalia Hispanica* 8, 1-131. [The vascular plants of Fuerteventura (Canary Islands), with special interest of forage. *Naturalia Hispanica* 8, 1-131].

López, J., & López, A. (1979). El clima de Canarias según la clasificación de Köppen. *Estudios Geográficos* 16(58), 5-21. [The climate of the Canary Islands according to the Köppen classification. *Geographic Studies* 16 (58), 5-21].

Martín, V. E., Scholz, S., & Wildpret de la Torre, W. (2009). Monitoring the invasive alien species on Fuerteventura (Canary Islands). In: P. Pyšek. & J. Pergl (Eds.), *Biological Invasions: Towards a Synthesis* (pp. 135–152). Prague: Neobiota 8.

Marzol, M. V. (1988). *La lluvia, un recurso natural para Canarias*. Sta. Cruz de Tenerife: Servicio de Publicaciones de la Caja General de Ahorros de Canarias. [*Rain, a natural resource for the Canary Islands*. Sta. Cruz de Tenerife: Publications Service of General Savings Bank of the Canary Islands].

Marzol, M. V. (1993). Los factores atmosféricos y geográficos que definen el clima de archipiélago canario. In *Aportaciones en homenaje al profesor Luis Miguel Albentosa* (151-176). Tarragona: Diputación de Tarragona. [The atmospheric and geographical factors that define the climate of the Canarian archipelago. In *Contributions in Homage to Professor Luis Miguel Albentosa* (151-176). Tarragona: Provincial Council of Tarragona].

Mcowen, C. J., Weatherdon, L. V., Van Bochove, J-W., Sullivan, E., Blyth, S., Zockler, C., Stanwell-Smith, D., Kingston, N., Martin, C. S., Spalding, M., & Fletcher, S. (2017). A global map of salt marshes. *Biodiversity Data Journal* 5: e11764. doi: 10.3897/BDJ.5.e11764.

Ministerio de Medio Ambiente. (2006). *Estudio Ecocartográfico del litoral de las islas de Fuerteventura y Lobos*. Madrid: Ediciones Tuomaï. [Ministry of the Environment (2006). *Eco-cartographic study of the coast of the islands of Fuerteventura and Lobos*. Madrid: Tuomaï Editions].

Mora, J. L., Arbelo, C. D., & Rodríguez, A. (2009). Características de los suelos de las Islas Canarias en relación a la vegetación natural. In E. Beltrán, J. Afonso-Carrillo, A. García & O. Rodríguez (Eds.), *Homenaje al Prof. Dr. Wolfredo Wildpret de la Torre* (pp. 665-684). La Laguna: Instituto de Estudios Canarios. [Characteristics of the soils of the Canary Islands in relation to natural vegetation. In E. Beltrán, J. Afonso-Carrillo, A. García & O. Rodríguez (Eds.), *Homage to Prof. Dr. Wolfredo Wildpret de la Torre* (pp. 665-684). La Laguna: Institute of Canarian Studies].

Morales, E. (2009). *Extracción de datos del Banco de Datos de Biodiversidad de Canarias para la evaluación de especies amenazadas. Ficha de Evaluación: Sarcocornia perennis. Actualización y análisis de los datos de las Especies Amenazadas de Canarias*. Colegio Oficial de Biólogos de Canarias-Viceconsejería de Medio Ambiente del Gobierno de Canarias. [*Extraction of data from the Canary Islands Biodiversity Data Bank for the evaluation of threatened species. Evaluation Sheet: Sarcocornia perennis. Updating and analysis of the data of the Threatened Species of the Canary Islands*. Official College of Biologists of the Canary Islands-Department of the Environment of the Canary Islands Government].

Pérez-Chacón, E., Hernández-Calvento, L., & Yanes, A. (2007). Transformaciones humanas y sus consecuencias sobre los litorales de las islas Canarias. In R. Paris & S. Etienne (Eds.), *Les littoraux volcaniques: une approche environnementale* (pp. 173-191). Clermont-Ferrand: Presses-universitaires Blaise Pascal. [Human transformations and their consequences on the littorals of the Canary Islands. In R. Paris & S. Etienne (Eds.), *Volcanic Coastlines: An Environmental Approach* (pp. 173-191). Clermont-Ferrand: University Press Blaise Pascal].

Pomel, R. S., Miallier, P., Fain, J., & Sanzelle, S. (1985). Datation d'un sol brun-rouge calcífere par une coulee d'âge Würm ancient (51.000 B.P.) à Fuerteventure (Iles Canaries). *Mediterranée* 4, 59-67. [Dating of calciferous brown-red soil by an age-old Würm (51,000 BP) melt at Fuerteventure (Canary Islands). *Mediterranean* 4, 59-67].

Pugnaire, F. I., Armas, C., & Valladares, F. (2004). Soil as a Mediator in PlantPlant Interactions in a Semi-Arid Community. *Journal of Vegetation Science* 15, 85-92.

Pye, K., & French, P. W. (1993). *Erosion and accretion processes on British salt marshes. Vol. I. Introduction: saltmarsh processes and morphology*. Cambridge: Environmental Research Consultants.

Rivas-Martínez, S., Díaz, T. E., Fernández-González, F., Izco, J., Loidi, J., Lousa, M., & Penas, A. (2002). Vascular plant communities of Spain and Portugal. *Itinera Geobotanica* 15 (1,2), 5-992.

Rodríguez, O., García, A., & Reyes, J. A. (2000). Estudio fitosociológico de la vegetación actual de Fuerteventura (Islas Canarias). *Vieraea: Folia scientarum biologicarum canariensium* 28, 61-98. [Phyto-sociological study of the current vegetation of Fuerteventura (Canary Islands). *Vieraea: Scientific Magazine of the Canary Biology* 28, 61-98].

Rodríguez, O. (Ed.). (2005). *Patrimonio natural de la isla de Fuerteventura*. Tenerife: Cabildo de Fuerteventura. Centro de la Cultura Popular Canaria. Consejería de Medio Ambiente y Ordenación Territorial. [*Natural heritage of the island of Fuerteventura*. Tenerife: The Island Council of Fuerteventura. Center for Canarian Popular Culture. Department of Environment and Territorial Planning].

Rodríguez, S., & Valdivia, C. (2009). *Evaluación de especies catalogadas de Canarias. Sarcocornia perennis. Expte Sarper 11/2009*. Gobierno de Canarias. Dirección General del Medio Natural, Servicio de Biodiversidad. [*Evaluation of cataloged species of the Canary Islands. Sarcocornia perennis. Expte Sarper 11/2009*. Government of the Canary Islands. General Directorate of the Natural Environment, Biodiversity Service]. Retrieved from the Government of the Canary Islands website: http://www.gobcan.es/opencmsweb/export/sites/medioambiente/piac/galerias/descargas/Documentos/Biodiversidad/evaluacion-especies/evaluacion-2009/Sarcocornia_perennis_09.pdf.

Romero, C. (1986). Aproximación a la sistemática de las estructuras volcánicas complejas de las Islas Canarias. *Ería. Revista de Geografía* 11, 211-223. [Approach to the systematics of the complex volcanic structures of the Canary Islands. *Ería. Journal of Geography* 11, 211-223].

Sadek, L. A., & Eldarier, S. M. (1995). Cyclic Vegetation Change and Pattern in a Community of *Arthrocnemum macrostachyum* in Mediterranean Coastal Desert. *Journal of arid Environment* 31, 67-76.

Santana, I., & Naranjo, M. (2002). *Sarcocornia perennis* (Miller) A. J. Scott. *Fuerteventura. Seguimiento de Poblaciones de Especies Amenazadas 2002*. Consejería de Medio Ambiente y Ordenación Territorial. [*Sarcocornia perennis* (Miller) A. J. Scott. *Fuerteventura. Monitoring of Endangered Species Populations* 2002. Department of Environment and Territorial Planning].

Zazo, C., Hansens, A., Goy, J. L., Criado, C., González-Delgado, J. A., Dabrio, C. J., & Ghaleb, B. (2008). Geomorfología volcánica y variaciones del nivel del mar en Lanzarote y Fuerteventura (Plio-Cuaternario). *Itinerarios Geológicos por las Islas Canarias. Lanzarote y Fuerteventura, Tenerife, La Gomera y El Hierro* 6, 143-172. [Volcanic geomorphology and sea level variations in Lanzarote and Fuerteventura (Plio-Quaternary). *Geological Itineraries in the Canary Islands. Lanzarote and Fuerteventura, Tenerife, La Gomera and El Hierro* 6, 143-172.].

In: Salt Marshes
Editors: D. Barnes and C. Ellis
ISBN: 978-1-53614-040-8

Chapter 2

ON THE SUBTROPICAL SALT MARSHES OF BRAZIL: DISCUSSIONS ABOUT A RECENTLY RECOGNIZED ECOSYSTEM

Mario Arthur Favretto*
Health Surveillance Department,
Municipal Government of Campos Novos,
Campos Novos, State of Santa Catarina, Brazil

ABSTRACT

The subtropical salt marshes of Brazil compose a recently recognized ecosystem (only in 2017) and are under human threat. This ecosystem is located in the South of Brazil, between the north-central coast of São Paulo state and the central coast of Santa Catarina state, has less than 20,000 ha and it was studied during many years by the researchers M. R. Bornschein and B. L. Reinert, resulting in being recognized as a new ecosystem. Despite several ornithological researches in these marshes, little is known about detailed aspects of their ecology and other groups of animals. Under strong diary tidal effects and sedimentation processes,

* Corresponding Author Email: marioarthur.favretto@hotmail.com.

although being currently suffering processes of tropicalization, sea level rising and exotic plant species invasions, it is a very dynamic ecosystem. The subtropical salt marshes of Brazil are extremely restrictive ecosystems to animals that live there, being under selection of different factors that result in reduced populations (at least to birds) in these marshes. These animals are frequently under influences of flooding by the tidals, adding up effects of predation, fall of vegetation by wind and influences of extreme climatic events. Therefore, the animal species are in a fragile trade-off to survive, that here I denominate, in homage to the researchers who first described these characteristics, ecological balance point of Borschein-Reinert. Different from variations in other ecosystem gradients, as latitudinal and altitudinal gradients. Vegetation variations in these marshes, which at first sight appear homogeneous, occur in a few meters. And, for instance, the nest site selection by birds results in strong influences over their reproductive success. The choice of a nest site few meters near or distant the water edge of the marshes or a few centimeters higher or lower, represents the difference between flooding or not flooding of the nest during the high tide. Consequently, the life or death of nestling. Therefore, the species occurring in this ecosystem are living in an extreme environment, always oscillating between environmental characteristics more or less benefic to it, and more successful will be those individuals able to live at these delicate ecological balance. Thus, here I will discuss a general description of this recently recognized ecosystem and the ecological influences over the species that live in these marshes, resulting in the fragile ecological balance point of Borschein-Reinert, and propose a mathematical index to this balance.

Keywords: Brazil, ecological balance, habitat restriction, marshes

INTRODUCTION

Salt and tidal marshes are areas with few occurrences of species, not in negative aspects, but related to the high restriction present in these environments (Greenberg, 2006). These characteristics are created by the high restrictions involved in the salt and tidal marshes, resulting in processes of sedimentation and erosion, tides influences, inundation and salinity (Watson & Byrne, 2009; Gedan et al., 2011). Due to these characteristics, a high species endemism degree in these marshes occurs, at least in North America, where marshes are best known; while in South America it is still very insipient this knowledge (Greenberg & Maldonado, 2006).

In South America, in particular in Brazil, there are many phytophysiognomies forming marshes in the coastal regions (Reinert et al., 2007). One of these was recently recognized as a new ecosystem because of its peculiarities, thus being called as subtropical salt marshes (previously, known as tidal marsh), covering coastal areas between the states of São Paulo and Santa Catarina, in the Southeast and South of Brazil (Bornschein et al., 2017).

Despite this recent recognition as such, the term "subtropical salt marshes" had already been used for certain local phytophysiognomies since the decade of 1990, but with more specificity to areas with *Spartina* spp. domain (Netto & Lana, 1997), this concept was modified by Bornschein et al. (2017). The peculiar characteristics of these marshes had been observed by Costa & Dajy (1992), which emphasize the influence of large rivers over estuarine lagoons and bays over coastal areas of Brazil where these marshes are located, with waters in brackish condition throughout the year and higher salinity as closer to the coast, and the domain of herbaceous vegetables species of Cyperaceae family.

Despite all the complexity of the interactions occurring in subtropical salt marshes, there are still few information related to its bioversity; the animal group better known in its areas are the birds. However, information on other animals groups are still scarce, and the same is true to plants and ecological interaction in this ecosystem. In addition, most of the information about these marshes appears only in thesis and dissertations not yet published and wrote in Portuguese.

Thus, in the present chapter I make a general analysis about the current knowledge of subtropical salt marshes in Brazil. I create the hypothesis of the occurrence of a new ecological concept that involves the fauna (especially the birds) that uses this ecosystem and propose an index for the analysis of populations of species that occur in the areas of these marshes.

Distribution and Plant Structure

The subtropical salt marshes (previously called tidal marshes) are located in estuarine regions inside bays, mainly in riverside of significant river input, where it prevents or hinders saline intrusion, in association with mangroves and with minimal occurrence of *Spartina* spp. (Bornschein, 2001; Reinert et al., 2007; Bornschein et al., 2017). These marshes have a distribution of less than 20.000 ha, between the north-central coast of São Paulo state and the central coast of Santa Catarina state (Bornschein et al., 2017).

From the central coast of the state of Santa Catarina heading south to Argentina, there is an area of transition from subtropical salt marshes to completely temperate influences (Costa & Dajy, 1992). To RADAMBRASIL project, as related by Bornschein (2001) and Reinert et al. (2007), the subtropical salt marshes are classified as Pioneering Formation of Fluvial-Marine Influence, denominated as tidal marshes in Reinert et al. (2007) and being recognized as a new ecosystem in Bornschein et al. (2017).

These marshes have domain of herbaceous pioneer vegetation, form some of the initial stages of plant succession in their areas of occurrence, and because of their location in estuarine areas, are under direct and indirect influences of the water level originated from tides variation and as a consequence of the salinity levels altered (Reinert et al., 2007; Bornschein, 2013; Bornschein et al., 2017).

There is vegetation development even where the presence of water is constant, plant height generally varies from 0.5 to 3 m, the richness of plant species varies from one to seventeen species per square meter, but 85 species have already been recorded in this ecosystem, belonging to 47 families, being: 62 herbaceous species, 12 trees, nine climbing plants and two shrubs (Reinert et al., 2007).

However there is the domain of some plant species: *Schoenoplectus californicus* (Cyperaceae), *Crinum americanum* (Amaryllidaceae), *Echinodorus grandiflorus* (Alismataceae), *Zizaniopsis microstachya*

(Poaceae), *Panicum* cf. *mertensii* (Poaceae) and *Brachiaria arrecta* (Poaceae; invasive alien species) (Bornschein, 2001; Reinert et al., 2007). There are also occasional groups with domain of *Cladium mariscus* subsp. *jamaicense* (Cyperaceae), *Typha domingensis* (Typhaceae), *Acrostichum danaeifolium* (Pteridaceae), shrubs of *Talipariti tiliaceum* var. *pernambucense*, besides some trees of *Laguncularia racemosa* (Combretaceae), *Calophyllum brasiliense* (Clusiaceae), *Annona glabra* (Annonaceae) and *Tabebuia cassinoides* (Bignoniaceae), with a lower prevalence of other plant species (Borschein, 2001; Reinert et al., 2007; Favretto et al., 2016a; Bornschein et al., 2017).

It should also be noted the reduced presence of *Spartina* spp. (Poaceae) in the subtropical salt marshes, whose presence would be considered common in salt marshes *lato sensu* (Bornschein et al., 2017). Species of this genus have a larger domain of the salt marshes in the southern latitudes, especially starting from the central area of the coast of Santa Catarina state, also in the state of Rio Grande do Sul, towards Argentina, where the climate has more temperate characteristics (Panitz, 1992, Isacch et al., 2006).

When thinking about marshes, wide dominated areas of almost homogeneous herbaceous vegetation can be imagined, similar to what occurs in some salt marshes of temperate regions, but in subtropical areas the situation is different. As mentioned above, there is a variable species richness of plants and also its structure is variable. In a subtropical salt marsh there is the formation of which we could consider different gradients of the vegetation structure; from water edge to inside the land there are small variations of altitude, salinity and sedimentation levels, these result also in variations in the density, height, grouping and composition of plant species (Reinert et al., 2007; Bornschein, 2013; Favretto et al., 2016a).

There is also variation in the structure of the subtropical salt marshes according to their larger or lesser proximity to the estuarine areas, that is, with a larger or lesser salinity effect. The vegetation structure, although herbaceous, but reaching almost 3 m, creates a plant stratification, which also exerts influences on the fauna of these marshes, facilitating or not the occurrence in some areas (Favretto et al., 2016a; Figures 1, 2 and 3).

Figure 1. View of the Subtropical Salt Marshes in Guaratuba bay, state of Paraná, South of Brazil. In the figure appears *Schoenoplectus californicus* (thinner and taller leaves) and *Crinum americanum* (smaller leaves). Photo by Mario Arthur Favretto.

Figure 2. View of the Subtropical Salt Marshes in Guaratuba bay, state of Paraná, South of Brazil. In the figure appears *Schoenoplectus californicus* (thinner and taller leaves) and *Crinum americanum* (smaller leaves and with white flowers), also a small mangrove (*Laguncularia racemosa*). Photo by Mario Arthur Favretto.

Figure 3. View of the Subtropical Salt Marshes in Guaratuba bay, state of Paraná, South of Brazil. In the figure appears *Schoenoplectus californicus* (thinner and taller leaves) and *Crinum americanum* (smaller leaves and with white flowers). Photo by Mario Arthur Favretto.

ANIMALS AND ECOLOGY

In the subtropical salt marshes, more specifically in their areas in the state of Paraná, south of Brazil, where there are studies with this ecosystem, there are 81 bird species recorded (Borschein, 2001), being the faunistic group better known in these marshes. There are also the record of 12 species of Lepidoptera: *Anartia amathea, Anartia jatrophae, Caligo* sp., *Danaus* sp., *Dryadula phaetusa, Dryas iulia, Junonia* sp., *Philaethria wernickei*, *Tegosa claudina*, *Morpho epistrophus*, *Battus polydamas* and *Heraclides thoas*; three species of Apidae (Hymenoptera): *Xylocopa* sp., *Apis mellifera* and *Trigona spinipes*; and nine species of Tabanidae (Diptera): *Acanthocera longicornis*, *Anaerythrops* cf. *lanei*, *Phaeotabanus* cf. *litigiosus*, *Chrysops* sp., three species of *Catachlorops* sp. and two species of *Tabanus* sp.

Many invertebrate species seems to use the subtropical salt marshes as opportunistic, considering that many species are not specialized in this ecosystem, being found in forests, forest edges or anthropized areas (Geuster, 2013; Favretto et al., 2015; Orlandin, 2017). In fact, the number of species effectively resident in salt or brackish marshes is usually low (Greenberg & Maldonado, 2006).

This opportunism is highlighted by the variation that occurs in the composition of insect species in the areas of the subtropical salt marshes, varying in relation to the proximity or not of areas with larger or less tidal effects (M.A. Favretto, obs.pess.). In this way, exposed to different levels of salinity (higher or lower), that could be one of the factors of influence on this group, as well as variations in floristic composition due to these factors.

About other invertebrates in these marshes, in areas with domain of *Schoenoplectus californicus* there are always a large abundance of insect of the family Delphacidae (Hemiptera) over these plants, these insects are food to birds of this ecosystem (Favretto et al., 2016a). Among other insects, there are, mainly in areas of edges with mangroves (*Laguncularia racemosa*) and "guanandizais" (*Calophyllum brasiliense*), records of the families: Curculionidae, Meloidae, Staphylinidae (Coleoptera); Diapriidae, Eulophidae, Ceraphroniidae, Scelionidae, Vespidae (Hymenoptera); Delphacidae, Corixidae, Gerridae (Hemiptera); Chironomidae, Culicidae (Diptera) (M.A. Favretto and E.B. dos Santos, obs. pess.).

However, about the animals of the subtropical salt marshes, the most known are the birds, in special Parana Antwren *Formicivora acutirostris* (Thamnophilidae), Many-colored Rush Tyrant *Tachuris rubrigastra* (Tachurisidae)and Wren-like Rushbird *Phleocryptes melanops* (Furnariidae). From these, the first is the better known about its biology in this ecosystem, although it was discovered only in the decade of 1990 (Bornschein et al., 1995; Reinert & Bornschein, 1996; Bornschein et al., 2015). With the biology data of these birds it is possible to observe that the occurrence of these species in subtropical salt marshes depends on different ecological adaptations for their survival in such an extreme environment, as I discuss below.

Figure 4. Wren-like Rushbird *Phleocryptes melanops* (Furnariidae). Photo by Mario Arthur Favretto.

In the case of *Formicivora acutirostris,* this bird is characterized by having territories inserted in the areas of subtropical salt marshes (not occurring within extensive forest areas). In *F. acutirostris* the larger territories are those that suffer with large areas of flooding, unlike individuals with smaller territories, then flooding induces individuals to have larger territories due to periods of high tide. Thus, there is compensation due to loss of area during periods of high tide (Corrêa, 2011).

In addition, *Formicivora acutirostris,* depending on the phyto-physiognomy, selects characteristics of its reproductive sites that vary, such as large height of vegetation in phytophysiognomy with domain of *Crinum americanum*. In the phytophysiognomy with domain of *Cladium mariscus* subsp. *jamaicense* large height of vegetation and large percentage of plant coverage; in areas with *Acrostichum danaefolium*,large plant heterogeneity and height. Avoiding areas with lower altitude and height of vegetation due to the possibility of flooding of nests during high tide (Reinert, 2008).

About the birds *Tachuris rubrigastra* and *Phleocryptes melanops* in these marshes, Favretto et al. (2016a) recorded that such birds occur in phytophysiognomies with domain by four herbaceous plant species (*Schoenoplectus californicus, Crinum americanum, Cladium mariscus* subsp. *jamaicense* and *Typha domingensis*). In these bird species the vegetation (plant coverage) is the variable with most influence in occurrence areas of these bird species. These areas are formed by a plant stratification, even if they are herbaceous plants. And the density of the vegetation in each stratum, being excessively or little dense, or vegetation excessively high or low, will have influence on the occurrence or not of the birds in these areas.

Depending on the phytophysiognomy with domain of some of the previous plant species, differences were recorded or not in the amount of exposed soil or in the amount of food resources for these birds, in this case, abundance of invertebrates, food source of these species (Favretto et al., 2016a). Also, differences in altitude of the areas, even of centimeters, in the marshes can represent the difference between a nest be flooded or not during a period of high tide (Picken & Delgado, 2014; Favretto et al., 2016a). In addition to influencing the fauna of vertebrates, the different structures of the vegetation of these marshes also exert a strong influence on the composition of the benthic fauna, generating variations in this as the vegetation density or sedimentation processes change (Netto & Lana, 1997).

The data recorded by Favretto et al. (2016a) was previously observed by Marcos R. Borschein and Bianca L. Reinert. Unifying these data with the present chapter is observable that in subtropical salt marshes, as in the example of *Formicivora acutirostris, Tachuris rubrigastra* and *Phleocryptes melanops*, there is an ecosystem with large restrictions, where the different phytosociological and ecological characteristics, as the availability of food resources and altitude, exert strong selective pressure over animal species. For example, to *F. acutirostris*, its simple rate of reproductive success from the incubation is less than 16% (Reinert, 2008). A success rate smaller than that of Passeriformes from the North hemisphere, data that demonstrates how extreme this ecosystem is, as in other similar marshes (Greenberg, 2006; Greenberg et al., 2006; Greenberg & Maldonado, 2006).

In these marshes, a lower altitude of only a few centimeters can represent the loss of nests and nestlings during high tide. Likewise, the greater proximity of the nesting site to the water edge, in contrast to a few meters or less in the nest location, may also result in areas with different densities and vegetation heights, and larger or lesser reproductive success of birds occurring on the area (Reinert, 2008; Bornschein, 2013; Favretto et al., 2016a).

In this way, all these constant searches and disputes for more adequate territorial conditions, successes and failures of the fauna, especially birds, in the subtropical salt marshes, is what I call the "ecological balance point of Bornschein-Reinert." Tribute to the researchers Marcos R. Borschein and Bianca L. Reinert, who, for several decades have studied the birds of this ecosystem, found new species of birds in their area and carried out studies that allowed here the classification of this ecological threshold and also the definition of these marshes with unique characteristics as an ecosystem in itself.

In the subtropical salt marshes the species are in a constant "fight" to survive, different from other ecosystems with periods of relative stasis or attenuation of selective pressures. Certainly there is in these marshes a strong pressure of natural selection. Besides the low reproductive success of *Formicivora acutirostris,* previously cited (Reinert, 2008), it should also be considered, for instance, that environmental restrictions influence even the seasonality and migration of some of the bird species of these marshes (Salewski & Bruderer 2007; Favretto et al., 2016b). As to *Tachuris rubrigastra* and *Phleocryptes melanops* that have a partial migration in two shifts with replacement of the individuals in two periods of larger abundance throughout the year (Favretto et al., 2016b).

Synthesizing the ecological balance point is also possible, by means of a mathematical index that I propose here. At the outset it is an initial tool for comparative purposes of the conservation status of populations of a species or of different areas of marsh (or other environment), considering that this can be adapted for the insertion of different ecological variables that are under analysis. That is, because it is a flexible index allows the insertion of different variables that can result in a trade-off for a given biological population.

As a proposal of this index, we have the sum of the natural logarithm of the ratio between data obtained for the population by the general data of the variable in the sample environment, multiplied by 0.1.

$$EB = \frac{\ln(a)}{\ln(b)} + \frac{\ln(c)}{\ln(d)} + \frac{\ln(e)}{\ln(f)} x\ 0.1$$

For example, if population A of a species of bird that inhabits the marshes at site A had 10 nests with reproductive success and two with losses, uses 90 m^2 of a total area of 100 m^2 of marshes, and the area of these marshes that suffers flood above 50 cm from the ground (assuming this could generate more nest losses) is 5 m^2 of 100 m^2 total. We will have:

$$EB(A) = \frac{\ln(10)}{\ln(2)} + \frac{\ln(90)}{\ln(100)} + \frac{\ln(5)}{\ln(100)} x\ 0.1$$

$$EB(A) = 0.46$$

While the B population of this species inhabiting the marshes in site B had two nests with reproductive success and 10 with losses, uses 5 m^2 of a total area of 100 m^2 of marshes, and the area of these marshes that suffers flooding of more than 50 cm of the soil (assuming this could generate more losses of nests) is 50 m^2 of 100 m^2 total. We will have:

$$EB(B) = \frac{\ln(2)}{\ln(10)} + \frac{\ln(5)}{\ln(100)} + \frac{\ln(50)}{\ln(100)} x\ 0.1$$

$$EB(B) = 0.15$$

In this way, it would be possible to infer that the population A of the species would be closer to an ecological balance point (EB = 0.46), in a sense, presenting a positive surplus balance. While population B would be in imbalance (EB = 0.15), with a tendency to undergo drastic population reductions or even local extinctions. Therefore, the proposed index can be used as a tool for species conservation, initially thought for birds in marshes areas, but it can be used for other ecosystems or phytophysiognomies, both for inter-population and temporal comparisons.

THREATS TO SUBTROPICAL SALT MARSHES

Despite the large complexity of these marshes and their recent recognition as an ecosystem, is long-established the threats that it has been already suffering. Among the main threats are the invasions of exotic species as well as climate change and sea level rising due to global warming, threats also detected in other estuarine and marshes areas (Takekawa et al., 2006; Adam, 2009; Leonardi et al., 2016; Hunter, 2017). The areas of these and other marshes in the region are being invaded by exotic grasses, *Hedychium coronarium, Brachiaria mutica* and *Brachiaria arrecta* (Reinert et al., 2007; Bornschein, 2013; Bornschein et al., 2017).

Occurring a large expansion of area used by the latter species in the last years, *B. arrecta* forms large agglomerations of plants that restrict the growth space of the native plants, also shading them. In addition, it forms banks of semi-floating plants that spread over the waters and which, when detached, increase the dispersion of the species in the marshes areas (Reinert et al., 2007; Bornschein, 2013). These areas of exotic grasses species do not only result in the destruction of native vegetation, but also prevent native birds from inhabiting these sites, reducing their areas of occurrence (Reinert et al., 2007; Favretto et al., 2016c). Facing with this environmental scenario, the biologists Marcos R. Borschein and Bianca L. Reinert have made enormous efforts to carry out the control and elimination of these exotic grasses that have invaded areas of subtropical salt marshes in the interior of Guaratuba bay, in the state of Paraná, southern Brazil.

The areas of these marshes will also suffer negative effects of climate change, such as sea level rising, changes in precipitation patterns and flood periods (Bornschein et al., 2017). All these may result in a significant loss of areas for the species living in these marshes. It is possible that these changes are already in progress in the subtropical salt marshes, Bornschein (2013) recorded changes in some marshes areas in the state of Paraná, with distribution expansion of *Panicum* cf. *mertensii* (native species) and also of the mangroves over marshes (tropicalization), as well as an increase in plant species richness (up to seven new species in some areas).

To Bornschein (2013), these changes are natural, in a view of the natural process of plant succession, but may also be the result of extreme climatic events recorded in their sample areas (floods and heavy rains). However, in some areas of the southern coast of Brazil (Paranaguá Bay, Paraná state), there is an increase in sedimentation due to deforestation in riparian areas above estuaries (Lana et al., 2001), remains to know if this sedimentation will help the expansion of marshes in order to delay part of the impacts of climate change (Adam, 2009).

CONCLUSION

The future of subtropical salt marshes is uncertain and its consolidation as an ecosystem as proposed by Bornschein et al. (2017) is essential for future research and conservation actions. Because these marshes are an ecosystem that is difficult to access and study, this has left them as a mystery in terms of a whole new range of ecological processes that still need to be analyzed or discovered, such as the ecological balance point of Bornschein-Reinert proposed here.

Several studies with other bird species, besides those mentioned here, as other groups of vertebrates, invertebrates and plants, still have to be made in these marshes, as well as the understanding of all physical-chemical and geological processes that imply the formation, expansion or retraction of its areas. However, all this complexity still unknown is under enormous threat of disappearing, without even being described. Extreme climatic events have already affected some areas of these marshes, and also occurred the invasion of exotic species and changes in the distribution of native species. With the intensification of global warming, there will still be sea level rising, coupled with tropicalization processes, which could completely uncharacterize this ecosystem and may even eliminate it completely.

ACKNOWLEDGMENTS

The author is thankful to Marcos R. Bornschein and Bianca Reinert by the opportunity to know the subtropical salt marshes, to Tiago Machado de Souza and Claudia Golec to their partnership in the marshes, and to Emili Bortolon dos Santos to review the chapter.

REFERENCES

Adam, P. (2009) Australian salt marshes in global context. In Saintilan, N. (ed.), *Australian Saltmarsh Ecology* (pp. 1-22). Australia: CSIRO.

Bornschein, M. R.; Reinert, B. L. & Teixeira, D. M. (1995) Um novo Formicariidae no sul do Brasil (Aves, Passeriformes). *Série Publicação Técnico-Científica Instituto Nova Iguaçu de Pesquisa e Preservação Ambiental,1,* 1-18. [A new Formicariidae in the south of Brazil (Aves, Passeriformes). *Scientific and Technical Publication Series Nova Iguaçu Institute for Research and Environmental Preservation, 1*, 1-18].

Bornschein, M. R. (2013) *Biologia da conservação do bicudinho-do-brejo Stymphalornis acutirostris (Aves, Thamnophilidae)*. Tese. Curitiba, PR: PPG Ecologia e Conservação/UFPR. [*Conservation biology of the Parana Antwren Stymphalornis acutirostis (Aves, Thamnophilidae)*. Thesis. Curitiba, PR: PPG Ecology and Conservation/UFPR].

Bornschein, M. R. (2001) *Formações pioneiras no litoral centro-sul do Paraná: identificação, quantificação de áreas e caracterização ornitofaunística*. Dissertação. Curitiba, PR: UFPR. [*Pioneer formations in central-south coast of Parana: identification, areas quantification and bird composition characterization*. Dissertation. Curitiba, PR: UFPR].

Bornschein, M. R.; Reinert, L. B.; Machado-de-Souza, T.; Golec, C.; Whitney, B. M. & Favretto, M. A. (2017) Abundance, occurrence, and seasonality of the Subtropical Doradito (*Pseudocolopteryx acutipennis*) on the coast of Brazil. *The Wilson Journal of Ornithology*, *129*, 199-206.

Bornschein, M. R.; Pizo, M. A.; Sobotka, D. D.; Belmonte-Lopes, R.; Golec, C.; Machado-de-Souza, T.; Pie, M. R. &Reinert, B. L. (2015) Longevity records and signs of aging in Marsh Artwren *Formicivora acutirostris* (Thamnophilidae). *The Wilson Journal of Ornithology*, *127*, 98-102.

Corrêa, L. (2011) *Influência da cobertura vegetal no tamanho de territórios de Stymphalornis acutirostris e análise da seleção de sítios de nidificação após insucesso reprodutivo*. Dissertação. Curitiba, PR: PPG Ecologia e Conservação/UFPR. [*Influence of plant cover on the territories size of Stymphalornis acutirostris and analysis of nesting site selection after reproductive failure*. Dissertation. Curitiba, PR: PPG Ecology and Conservation/UFPR].

Costa, C. S. B. & Davy, A. J. (1992) Coastal saltmarsh communities of Latin America. In Seelinger, U. (Ed.), *Coastal plant communities of Latin America* (pp. 179-199). California: Academic Press.

Favretto, M. A. (2016) *Restrição a locais de reprodução e sazonalidade de Tachuris rubrigastra (Aves: Tachurisidae) e Phleocryptes melanops (Aves: Furnariidae) no litoral sul do Brasil*. Dissertação. Curitiba, PR: PPG Ecologia e Conservação/UFPR. [*Reproductive sites restriction and seasonality of Tachuris rubrigastra (Aves: Tachurisidae) and Phleocryptes melanops (Aves: Furnariidae) in the coast of the south of Brazil*. Dissertation. Curitiba, PR: PPG Ecology and Conservation/UFPR].

Favretto, M. A.; Piovesan, M.; Orlandin, E. & Santos, E. B. (2015) Lepidoptera em um fragmento florestal urbano no sul do Brasil.

Scientia Plena, 11, 1-6. [Lepidoptera in an urban forest remnant in the south of Brazil. *Scientia Plena, 11*, 1-6].

Favretto, M. A.; Machado-de-Souza, T.; Golec, C.; Reinert, B. L. & Borschein, M. R. (2016a) Seleção de habitat por *Tachuris rubrigastra* (Tachurisidae) e *Phleocryptes melanops* (Furnariidae) em brejos salinos subtropicais, Brasil. InFavretto, M. A. *Restrição a locais de reprodução e sazonalidade de Tachuris rubrigastra (Aves: Tachurisidae) e Phleocryptes melanops (Aves: Furnariidae) no litoral sul do Brasil* (pp. 49-106). Dissertação. Curitiba, PR: PPG Ecologia e Conservação/UFPR. [Habitat selection in *Tachuris rubrigastra* (Tachurisidae) and *Phleocryptes melanops* (Furnariidae) in subtropical salt marshes, Brazil. In Favretto, M. A. *Reproductive sites restriction and seasonality of Tachuris rubrigastra (Aves: Tachurisidae) and Phleocryptes melanops (Aves: Furnariidae) in the coast of the south of Brazil.* Dissertation. Curitiba, PR: PPG Ecology and Conservation/UFPR].

Favretto, M. A.; Machado-de-Souza, T.; Golec, C.; Reinert, B. L. & Borschein, M. R. (2016b) Sazonalidade de *Tachuris rubrigastra* (Tachurisidae) e *Phleocryptes melanops* (Furnariidae) em brejos salinos subtropicais no sul do Brasil. InFavretto, M.A. 2016. *Restrição a locais de reprodução e sazonalidade de Tachuris rubrigastra (Aves: Tachurisidae) e Phleocryptes melanops (Aves: Furnariidae) no litoral sul do Brasil* (pp. 107-138). Dissertação. Curitiba, PR: PPG Ecologia e Conservação/UFPR. [Seasonality of *Tachuris rubrigastra* (Tachurisidae) and *Phleocryptes melanops* (Furnariidae) in subtropical salt marshes in the south of Brazil. In Favretto, M. A. *Reproductive sites restriction and seasonality of Tachuris rubrigastra (Aves: Tachurisidae) and Phleocryptes melanops (Aves: Furnariidae) in the coast of the south of Brazil* (pp. 107-138). Dissertation. Curitiba, PR: PPG Ecology and Conservation/UFPR].

Favretto, M. A.; Reinert, B. L.; Machado-de-Souza, T.; Golec, C. & Bornschein, M. R. (2016c) Status de ocorrência de *Tachuris rubrigastra* (Tachurisidae) e *Phleocryptes melanops* (Furnariidae) em brejos subtropicais no Brasil e extrema restrição a áreas de nidificação.

InFavretto, M. A. *Restrição a locais de reprodução e sazonalidade de Tachuris rubrigastra (Aves: Tachurisidae) e Phleocryptes melanops (Aves: Furnariidae) no litoral sul do Brasil* (pp. 19-48). Dissertação. Curitiba, PR: PPG Ecologia e Conservação/UFPR. [Status of occurrence of *Tachuris rubrigastra* (Tachurisidae) and *Phleocryptes melanops* (Furnariidae) in subtropical marshes of Brazil and extreme restriction in nesting sites. In Favretto, M. A. *Reproductive sites restriction and seasonality of Tachuris rubrigastra (Aves: Tachurisidae) and Phleocryptes melanops (Aves: Furnariidae) in the coast of the south of Brazil* (pp. 19-48). Dissertation. Curitiba, PR: PPG Ecology and Conservation/UFPR].

Gedan, K. B.; Altieri, A. H. & Bertness, M. D. (2011) Uncertain future of New England salt marshes. *Marine Ecology Progress Series*, *434*, 229-237.

Geuster, C. (2013) Abelhas sociais sem-ferrão do meio oeste de Santa Catarina. InFavretto, M. A. & E. B. Santos (Eds.), *Insetos do oeste de Santa Catarina* (pp. 164-227). Campos Novos, SC: Mario Arthur Favretto. [Stingless bess in the West of Santa Catarina. In Favretto, M. A. & E. B. Santos (Eds.), *Insects of the west of Santa Catarina* (pp. 164-227). Campos Novos, SC: Mario Arthur Favretto].

Greenberg, R. (2006) Tidal marshes: home for the few and the highly selected. *Studies in Avian Biology*, *32*, 2-9.

Greenberg, R. & Maldonado, J. E. (2006) Diversity and endemism in tidal-marsh vertebrates. *Studies in Avian Biology*, *32*, 32-51.

Greenberg, R.; Elphick, C.; Nordby, J. C.; Gjerdrum, C.; Spautz, H.; Shriver, G.; Schmeling, B.; Olsen, B.; Marra, P.; Nur, N. & Winter, M. (2006) Flooding and predation: trade-offs in the nesting ecology of Tidal-Marsh Sparrows. *Studies in Avian Biology*, *32*, 96-109.

Hunter, E. A. (2017) How will sea-level rise affect threats to nesting success for Seaside Sparrow? *The Condor*, *119*, 459-468.

Isacch, J. P.; Costa, C. S. B.; Rodríguez-Gallego, L.; Conde, D.; Escapa, M.; Gagliardini, D. A. & Iribarne, O. O. (2006) Distribution of saltmarsh plant communities associated with environmental factors along a latitudinal gradient on the south-west Atlantic coast. *Journal of Biogeography*, *33*, 888-900.

Lana, P. C.; Marone, E.; Lopes, R. M. & Machado, E. C. (2001) The subtropical estuarine complexo of Paranaguá Bay, Brazil. In Seelinger, U. & B. Kjerfve (Eds.), *Coastal marine ecosystems of Latin America* (pp. 131-145). Berlin: Springer.

Leonardi, N.; Ganju, N. K. & Fagherazzi, S. (2016) A linear relationship between wave power and erosion determines salt marsh resilience to violent storms and hurricanes. *Proceedings of the National Academy of Sciences*, *113*, 64-68.

Netto, S. A. & Lana, P. C. (1997) Intertidal zonation of benthic macrofauna in a subtropical salt marsh and nearby unvegetated flat (SE, Brazil). *Hydrobiologia*, *353*, 171-180.

Orlandin, E. (2017) *Tamanho do fragmento e microhabitats determinam riqueza e composição das espécies de borboletas em Floresta Atlântica*. Trabalho de Conclusão de Curso. Joaçaba, SC: Unoesc. [*Remnant size and microhabitat determine species richness and composition of butterflies in Atlantic Rainforest*. Joaçaba, SC: Unoesc.].

Pantiz, C. M. N. (1992) Ecological aspects of a saltmarsh ecosystem in Santa Catarina Island, Brazil. In Seelinger, U. (Ed.), *Coastal plant communities of Latin America* (pp. 213-230). California: Academic Press.

Pickens, B. A. & King, S. L. (2014) Multiscale habitat selection of wetland birds in the northern Gulf Coast. *Estuaries and Coasts*, *37*, 1301-1311.

Reinert, B. L. & Bornschein, M. R. (1996) Descrição do macho adulto de *Stymphalornis acutirostris* (Aves: Formicariidae). *Ararajuba*, *4*, 103-105. [Description of the adult male of *Stymphalornis acutirostris* (Aves: Formicariidae). *Ararajuba*, 4, 103-105].

Reinert, B. L. Bornschein, M. R. & Firkowski, C. (2007) Distribuição, tamanho populacional, hábitat e conservação do bicudinho-do-brejo *Stymphalornis acutirostris* Bornschein, Reinert e Teixeira, 1995 (Thamnophilidae). *Revista Brasileira de Ornitologia*, *15*, 493-519. [Distribution, population size, habitat and conservation of Parana

Antwren *Stymphalornis acutirostris* Bornschein, Reinert e Teixeira, 1995 (Thamnophilidae). *Brazilian Journal of Ornithology*, *15*, 493-519.]

Reinert, B. L. 2008. *Ecologia e comportamento do bicudinho-do-brejo (Stymphalornis acutirostris Borschein, Reinert & Teixeira, 1995 – Aves, Thamnophilidae)*. Tese. Rio Claro, SP: Unesp. [*Ecology and behavior of Parana Antwren (Stymphalornis acutirostris Borschein, Reinert& Teixeira, 1995 – Aves, Thamnophilidae)*. Thesis. Rio Claro, SP: Unesp.].

Saleweski, V. & Bruderer, B. (2007) The evolution of bird migration - a synthesis. *Naturwissenschaften*, *94*, 268-279.

Takekawa, J. Y.; Woo, I.; Spautz, H.; Nur, N.; Grenier, J. L.; Malamud-Roam, K.; Nordby, J. C.; Cohen, A. N.; Malamud-Roam, F. & Wainwright-de la Cruz, S. E. (2006) Environmental threats to tidal-marsh vertebrates of the San Francisco Bay estuary. *Studies in Avian Biology*, 32, 176-197.

Watson, E. B. & Byrne, R. (2009) Abundance and diversity of tidal marsh plants along the salinity gradient of the San Francisco Estuary: implications for global change ecology. *Plant Ecology*, *205*, 113-128.

Biographical Sketch

Mario Arthur Favretto

Affiliation: Municipal Government of Campos Novos, Campos Novos, state of Santa Catarina, Brazil

Education: MSc. in Ecology and Conservation

Research and Professional Experience: Mario Arthur Favretto is biologist, with master degree in ecology and conservation, has experience in research with Brazilian birds, mammals, amphibians, reptiles, orchids, bromeliads and insects. In the areas of ornithology, herpetology, entomology and ecology has been made researches for more than 10 years. Mario Arthur Favretto is author of more than 40 papers, and author, co-author or editor of eight books in these areas of biology. And using the alias "Marius Arthorius" published seven books of fiction, literature and poems.

Honors: Two honorable mentions by researches with ornithology and herpetology, in the VI Scientific Initiation Seminar and V Research Seminar of the University of the West of Santa Catarina.

Publications from the Last 3 Years:

Papers:

Santos, E. B.; Silva, P. G.; Favretto, M. A.; Muller, G. A. 2017. Occurrence of beetles (insecta: Coleoptera) in mosquito ovitraps (Diptera: Culicidae, Culicinae) in an Atlantic Forest fragment. *Revista Brasileira de Zoociências*, 18: 41-46. [*Brazilian Journal of Zoosciences*]

Bornschein, M. R.; Reinert, B. L.; Machado-De-Souza, T.; Golec, C.; Whitney, B. M.; Favretto, M. A. 2017. Abundance, occurrence, and seasonality of the Subtropical Doradito (*Pseudocolopteryx acutipennis*) on the coast of Brazil. *Wilson Journal of Ornithology*, 129: 199-206.

Costa, S. G. S.; Klompen, H.; Santos, E. B.; Favretto, M. A.; Pepato, A. R. 2017. Two new Brazilian Parasitengona larvae, Callidosoma (Acari, Erythraeidae) parasite of Lepidoptera and Durenia (Acari: Trombellidae) parasite of Culicidae (Diptera) with keys to the species. *Systematic and Applied Acarology*, 22: 42-57.

Favretto, M. A. 2017. Teoria neutra de biodiversidade: controvérsias e uma transvaloração da conservação de espécies. *Neotropical Biology and Conservation*, 12: 224-231. [Neutral Theory of Biodiversity: Controversies and a Transvaluation of Species Conservation]

Favretto, M. A.; Santos, E. B. 2017. New Records of Lepidoptera (Insecta) in Altitudinal Grasslands in State of Santa Catarina, Southern Brazil. *Acta Biológica Catarinense*, 4: 62-67.

Favretto, M. A.; Orlandin, E.; Santos, E. B.; Piovesan, M.; Onghero Jr., O. 2017. Lepidoptera composition variation in riparian forest in the south of Brazil. *Biota Amazônia*, 7: 26-29.

Santos, E. B.; Orlandin, E.; Piovesan, M.;Favretto, M. A. 2016. Nota sobre os mosquitos (Diptera: Culicidae) de um fragmento florestal urbano de Joaçaba, Santa Catarina, Brasil. *Entomotropica*, 31: 91-94. [Note on mosquitoes (Diptera: Culicidae) from an urban forest fragment of Joaçaba, Santa Catarina, Brazil.]

Orlandin, E.; Santos, E. B.; Piovesan, M.; Favretto, M. A.; Schneeberger, A. H.; Oliveira, V. S.; Muller, G. A.; Wagner, G. 2016. Mosquitoes (Diptera: Culicidae) from crepuscular period in an Atlantic Forest area in Southern Brazil. *Brazilian Journal of Biology*, 77: 1-8.

Dos Santos, E. B.; Favretto, M. A.; Dos Santos Costa, S. G.; Navarro-Silva, M. A. 2016.Mites (Acari: Trombidiformes) parasitizing mosquitoes (Diptera: Culicidae) in an Atlantic Forest area in southern Brazil with a new mite genus country record. *Experimental & Applied Acarology*, 69: 323-333.

Favretto, M. A.; Piovesan, M.; Orlandin, E.; Santos, E. B. 2015. Lepidoptera em um fragmento florestal urbano no sul do Brasil. Scientia Plena, 11: 1-6. [Lepidoptera in an urban forest fragment in southern Brazil. *Scientia Plena*]

Favretto, M. A. 2015. Estrutura da avifauna em um fragmento florestal no norte do Rio Grande do Sul, Brasil. *Neotropical Biology and Conservation*, 10: 132-142. [Structure of the avifauna in a forest fragment in the north of Rio Grande do Sul, Brazil.]

Favretto, M. A. 2015. Comparação entre a avifauna de três remanescentes florestais urbanos e um parque natural no sul do Brasil. Atualidades Ornitológicas, 185: 33-39. [Comparison between the avifauna of three urban forest remnants and a natural park in southern Brazil. *Ornithological News*]

Guzzi, A.; Gomes, D. N.; Santos, A. G. S.; Favretto, M. A.; Soares, L. M. S.; Carvalho, R. A. V. 2015. Composição e dinâmica da avifauna da usina eólica da praia da Pedra do Sal, Delta do Parnaíba, Piauí, Brasil. Iheringia. Série Zoologia, 105: 164-173. [Composition and dynamics of the avifauna of the Pedra do Sal beach wind farm, Parnaíba Delta, Piauí, Brazil. Iheringia. *Zoology Series*]

Orlandin, E.; Piovesan, M.; Favretto, M. A.; D'agostini, F.M. 2015. Mamíferos de médio e grande porte atropelados no oeste de Santa Catarina, sul do Brasil. *Biota Amazônia*, 5: 125-130. [Medium and large mammals run over in the west of Santa Catarina, southern Brazil.]

Books:

Favretto, M. A.; Geuster, C. J. 2017. Orquídeas e bromélias do oeste de Santa Catarina, Brasil. Campos Novos: Mario Arthur Favretto. 259 p. [*Orchids and bromeliads from the west of Santa Catarina, Brazil*]

Orlandin, E.; Favretto, M. A.; Piovesan, M.; Santos, E. B. 2016. Borboletas e mariposas de Santa Catarina: uma introdução. Campos Novos: Mario Arthur Favretto. 213 p. [*Butterflies and moths of Santa Catarina: an introduction.*]

In: Salt Marshes
Editors: D. Barnes and C. Ellis
ISBN: 978-1-53614-040-8

Chapter 3

SALT MARSH FAUNAL COMMUNITIES

Daiane Carrasco**, PhD
Instituto de Ciências Biológicas – ICB,
Universidade Federal do Rio Grande – FURG,
Rio Grande, State of Rio Grande do Sul, Brazil

ABSTRACT

Salt marshes develop in estuaries where there is reduced wave action and which is a source of sediment and suitable conditions for marsh plants to grow. The vegetation generally provides a structural habitat on featureless soft-sediment bottoms, and so it is utilized by a wide and diverse range of fish and invertebrates as their physical home, food supply, and shelter from predators. Salt marshes host a unique mix of marine and terrestrial animals including insects, birds, fish, and particularly crabs, whose activities heavily influence the productivity of salt marshes. Most of the primary production and detritus derived from marsh plants are trapped by plants and consumed by terrestrial organisms. Species such as insects and crabs demonstrate variable levels of adaptation to life in this intertidal habitat. Species composition and abundance of benthic invertebrates are strongly influenced by changes in the vegetation. High marshes, made up of shrubs and trees, provide a

* Corresponding Author Email: daiane_carrasco@hotmail.com.

tridimensional physical structure that supports a considerable biodiversity of such associated animals as insects, small mammals, birds, and reptiles.

A Changing Habitat

Salt marshes are among the most productive systems in the world (Begon et al., 1996). They are intertidal ecosystems located at the interface of marine and terrestrial systems. Salt marshes have strong directional and predictable physical and ecological gradients and a limited species pool of specialist fauna and flora (Adam, 1990).

Because salt marshes are close to marine areas and subject to flooding, salinity is, in fact, the major determinant of the distribution of salt marshes. Where salinity is above five, the plants are mostly salt-tolerant and are called halophytes (Figure 1). Where salinity is less than five, halophytes are mostly replaced by freshwater plants. Thus, salt marshes are replaced by other species at the freshwater ends of estuaries. This replacement is due more to competition with freshwater plants than to any physiological inability to withstand fresh water. At the very top of the shore, where tides no longer penetrate, halophytic vegetation is normally replaced by terrestrial vegetation (Little, 2000).

Figure 1. The halophyte vegetation is characteristic of salt marshes. *Spartina densiflora* is one of the salt marsh building plants and it forms circular tufts. The variation of salinity is a determinant factor in this ecosystem, due to its proximity to the sea and because there are several tidal channels that exchange between freshwater and salt water.

The plant community of coastal salt marshes is composed of relatively few species because they must have architectural and metabolic adaptations to survive in waterlogged soils (Chapman, 1974; Costa & Davy, 1992). Most salt marsh plants have adventitious roots near the soil surface to facilitate root oxygenation. Therefore, even with low richness, vascular marsh plants modify the physical structure of the soil and the amount and quality of light reaching the sediment, thus affecting temperature (Bertness & Hacker, 1994) and algal growth (Seliskar et al., 2002). These changes in key abiotic environmental factors appear to induce changes in the sediment biotic community (Whitcraft & Levin, 2007). On a large scale, critical salt marsh functions, such as nursery habitat provision, coastal stabilization, runoff filtration, and trophic support, are directly and indirectly due to the presence of vascular plants (Gleason et al., 1979; Warren & Neiring, 1993).

Studies in Brazilian marshes have identified changes in macrobenthos associated with plant biomass, detrital input, grain size, predation pressure, sediment organic matter, and freshwater input (Lana & Guiss, 1992; Pagliosa & Lana, 2005). Schrama et al., (2012) followed the succession of the invertebrate assembly in a European salt marsh for a time series of 100 years. They described that at the beginning of the succession incoming fluxes of organic matter and decomposers are mostly marine, whereas the importance of terrestrial matter and invertebrates increases over time.

This happens because salt marsh dynamics is driven by the frequency of tidal inundation and sediment supply. When salt marshes are flooded, estuary water brings fine sediments and rich nutrients, and deposits them in vegetation because the plants reduce wave energy in shallow coves. This natural fertilization is so effective that plant growth in salt marshes far outweighs the production of such cultivated plants as sugarcane (Seeliger et al., 2004). When there is a large accumulation of sediments, vegetation communities respond with ecological succession (Adam, 1990; Figure 2). Without the impact of herbivores, salt marshes move towards typical late successional stages over time (Kuijper et al., 2004).

Figure 2. Salt marshes are transitional environments. Over time, the accumulation of sediments creates a slope and terrestrial characteristics evolve in their highest part. Thus, small shrubs gradually replace grasses.

HERBIVORES

Salt marshes shelter a diverse assemblage of herbivores that at times can strongly reduce plant biomass and/or mediate plant distributions (Pennings & Bertness, 2001). The levels of herbivore damage are commonly high enough to have a major impact on plant survival and seed production (Rand, 2002).

Figure 3. Agglomerations of insects are common in the vegetation. Dipterans are frequent and abundant in salt marshes. In the picture we can see Calliphoridae flies exploring a carcass. They are important decomposers in this ecosystem. In the lowest part of marshes, crabs stand out. *Neohelice granulata* is an important herbivore of South West Atlantic salt marshes.

The consumers can be important determinants of plant distribution and abundance across environmental gradients (Figure 3; Louda & Rodman, 1996). Insects such as some hemipteran species could be important consumers, since they suck out the plant juices.

Insects

Insects, mainly phytophagous species, are conspicuous inhabitants of salt marshes. Sapsuckers such as aphids and nectar-feeders such as butterflies and dipterans are abundant among and above the stems of the grasses and herbs (Little, 2000). Salt marshes also have small shrubs and ferns that restrict the flow of water, this way favoring the development of insect larvae, mainly mosquitoes. The shrub vegetation offers shelter to a rich assemblage of insects that find favorable conditions for their reproduction and growth (Seeliger et al., 2004). A great diversity of spiders is also found in the plant canopy of most marshes, ranging from mobile hunters to sedentary web spinners (Little, 2000).

In American marshes, some of the dipterans are the first to make your acquaintance: while the males of horse flies (*Tabanus*), deer flies (*Chrysops*), and mosquitoes (*Aedes*) feed on nectar and pollen, females are vicious biters. Diptera could represent 40% of the total richness of insects (Hobach et al., 1999).

Diptera have a great capacity of dispersion, selection of environments and hosts; furthermore, its population dynamics is highly influenced by heterogeneity within the same habitat (Hobach et al., 1999). They are also important because of their role in ecosystems as predators, parasites, phytophagous, saprophagous, and pollinators (Wirth & Stone, 1968; Boer, 1981; Rosenberg et al., 1986). In particular, suborder Nematocera (a group of mosquitoes) is holometabolous, i.e., their larvae are mostly aquatic (Merritt et al., 2008), which makes the insects of this order biological bridges between aquatic and terrestrial trophic chains (Pupedis, 1997).

Mosquitoes are rife and may be accompanied by the biting midges or Ceratopogonidae (Little, 2000). Biting midges of genus *Culicoides* are found in salt marshes around the world (Forattini et al., 1958; Kline & Axtell, 1976). They cause serious disturbance to humans and animals because they act as disease vectors, including those cause by viruses, protozoa, and nematodes (Spinelli &Wirth, 1993).

In salt marshes, some beetles have an interesting life strategy. The staphylinid *Bledius spectabilis* frequently has large populations at the seaward edge of European salt marshes. *Bledius* not only lives in the burrows, the female lays her eggs there and the young hatch and grow up in the burrow before dispersing (Wyatt, 1986). The female *Bledius* also defends the burrow against predators, such as the carabid beetle, *Dicheirotrichus gustavi*. This predator has no burrow and retreats to cracks in the mud during the day. It has an endogenous circadian rhythm, which triggers emergence at night, and it then preys on *Bledius* eggs and larvae. When female *Bledius* are experimentally removed from burrows, *Dicheirotrichus* may eat the entire brood and then take over the burrow. Even if *Dicheirotrichus* does not take over, however, the burrows become blocked by sediment and eggs often become infected by fungus, so survival is low. *Bledius* is not alone in its habitats: several other species of salt marsh beetles look after their brood to some degree, showing that parental care by the female is essential in a habitat that can be regarded as harsh – at least for animals of terrestrial origin (Little, 2000).

BIOTURBATORS

Benthic macrofauna increase the O_2 penetration depths and redox conditions in sediments and stimulate microbial metabolism through feeding, burrowing, and burrow irrigation (Furukawa et al., 2001).

In the extensive salt marshes along the Atlantic Coast of the American continent, from Newfoundland (Canada) to southern Argentina, the smooth cordgrass *Spartina alterniflora* is a predominant vascular plant long believed to contribute considerably to energy flow in tidal-marsh estuaries (Peterson & Howarth, 1987; Silliman & Bertness, 2002). Although the gut contents of most mobile consumer-species often include large quantities of vascular plant detritus, there is little evidence that a strict diet of this material can sustain the populations of higher-order consumer-species, such as fishes (Kneib, 1997). However, microbial decomposers in the salt marshes of *S. alterniflora* leaves can play a key role in the trophic biology of some prominent lower-level consumers in the salt marshes (Kneib et al., 1997) and thus contribute to the maintenance of a prey pool for higher-order consumers (Graça et al., 2000). Most of these salt marshes are highly bioturbated by the burrowing crab *Neohelice* (= *Chasmagnathus*) *granulata* (Iribarne et al., 1997).

Crabs

Crab beds of *Neohelice granulata* constitute a conspicuous characteristic of South West Atlantic coastal areas from tidal flats to salt marshes. Crab density in the salt marsh is variable both within and among salt marshes ranging from six to more than 120 burrows per m^2 (Alberti et al., 2007). There is usually one crab and one entrance per burrow (Iribarne et al., 1997). As a bioturbator, this crab can remove up to 2.4 kg of sediment per day per m^2 and their burrows can reach up to 1 m depth and range from 2 to 7.5 cm in entrance diameter (Iribarne et al., 1997). The impacts on ecological functions in these salt marshes are either via herbivory, consuming large amounts of plant biomass, or via bioturbation, through the construction and maintenance of their burrows (Alberti et al., 2015).

Although there are other burrowers, such as the polychaetes *Laeonereis culveri* in Argentinean salt marshes (Albano et al., 2012), *Nephtys fluviatilis* and the amphipod *Kalliapseudes schubartii* in Brazil (Bemvenuti, 1992), herbivore organisms (stem-borer moths (Canepuccia et al., 2010), and rodents in the upper marsh (Pascual et al., 2015) inhabiting salt marshes, none have effects comparable to those of *N. granulata* on salt marsh ecological functioning (Alberti et al., 2015). Rodents can have a large influence via herbivory by reducing plant biomass, but these are restricted to the upper marsh (Pascual et al., 2015) and they inhabit terrestrial vegetation, thus play no role as burrowers in the salt marsh. *Cryptograpsus angulatus* is a grapsid crab that builds burrows in salt marshes in Caleta Valdés, at the southern limit of *N. granulata* distribution in 42°15'S (Iribarne et al., 2003). However, when both species coexist in northern coastal areas, *C. angulatus* is limited to inhabit the low intertidal and it never inhabits salt marsh areas (Martinetto et al., 2007). *C. angulatus* burrows are similar in shape to those of *N. granulata*, and it has been proposed that these species play similar role modifying sediments structure and modifying sediment structure and affecting the infaunal community in soft-bottom intertidal areas (Martinetto et al., 2011).

Crab burrows increase soil oxygenation promoting aerobic decomposition of the organic matter in the sediment and affecting the architecture and the production of roots (Daleo & Iribarne, 2009). The increment of O_2 in the sediment promotes arbuscular mycorrhizal fungi association, which in turn increases the productivity of the salt marsh. Plants growing in soils with a greater number of burrows have better quality and are preferred by herbivores (Martinetto et al., 2016).

Annelids

Oligochaetes can represent a major fraction of the benthic invertebrate fauna (Jackson et al., 1985), but this could change during the stage of succession in salt marshes. Levin & Talley (2002) compared a restored and an adjacent natural wetland system in Southern California. They observed that during early succession, when the marsh had little plant cover, the macrofaunal assemblage had a lower proportion of oligochaetes and a higher proportion of insect larvae as compared to the assemblage in the neighboring mature marsh. As the vegetation expanded and the created mash matured, the percentage of insect larvae decreased, and the percentage of polychaetes and amphipods increased.

Healy & Walters (1994) studied Oligochaeta assemblage in Sapelo Island, USA. They showed that species of the same genus (*Marionina*:

Enchytraeidae) have distinct preference living associated with *Spartina alterniflora*. *M. spartinae* was predominant in dead and live stems. It was the only species living inside the leaf sheaths, and was more abundant above 10 cm in high and low marsh and above 20 cm in creekside dead stems. *Marionina waltersi* was rare in high and low marsh samples and was associated mainly with the later stages of sheath decomposition. On creekside dead stems, *M. waltersi* was found 50 cm above the marsh surface. The distribution of both *M. spartinae* and *M. waltersi* on dead stems above 10 cm suggested a limited dependence on wetting stems during normal semidiurnal tides. *Marionina paludis* was present on dead stems in all marsh sites, but was only found on creekside live stems.

The presence of oligochaetes in the above ground parts of *Spartina* plants has two obvious explanations: plant stems represent a more favorable physical environment than the marsh sediments and/or a refuge from predation. Conditions within marsh sediments are usually reducing, with an absence of free oxygen and high concentrations of potentially toxic compounds such as H_2S, Mn^{2+} and Fe^{2+} (Oremland et al., 1982). Oligochaetes are sensitive to anoxic conditions (Giere & Pfannkuche, 1982), and meiofaunal organisms in marsh sediments typically are distributed within the first few millimeters of the sediment surface (Sacco et al., 1994).

Densities of worms in low marsh sediments are relatively low, and worms are found only within the surface sediments, less than 1 cm deep (Healy & Walters, 1994). Oxygen concentrations can be greater around *Spartina* roots where diffusion creates an oxygen 'halo' in the rizosphere (Teal & Kanwisher, 1961), but not enough oxygen may be released for respiration in saturated sediments, especially during tidal flooding (Gleason & Zieman, 1981). Few oligochaetes or other infauna were found associated with the roots and rhizomes, indicating that these animals are not able to avail of the increased oxygen concentration in the rhizosphere (Healy & Walters, 1994). Stem habitats probably do not suffer from the same limited availability of oxygen that is common in sedimentary habitats. Within the aerenchyma system of live *Spartina* plants, O_2 and

CO_2 gradients exist created by photosynthesis and the flux of gases between leaves and roots (Teal & Kanswisher, 1966). The concentration of both gases varies according to diurnal and tidal cycles (Gleason & Zieman, 1981).

Worms inhabiting the first millimeters of marsh sediment surface might avoid anoxic conditions, but they are exposed to predation from the many fish and crustaceans, which move over the marsh and forage during high tide (Rozas & LaSalle, 1990). Fauna found in live or dead stems presumably are less vulnerable to predation from nekton, but stem oligochaetes would also be vulnerable to predation by other dwelling fauna (Healy & Walters, 1994).

In the case of polychaetes, habitat specialists may be more susceptible to plant changes than habitat generalists, while the latter may be more affected by the impact of sediment disturbance in Brazilian salt marshes than the former (Pagliosa & Lana, 2005). Habitat generalists, such as the numerically dominant deposit-feeder polychaete *Capitella* sp. and *Laeonereis acuta*, colonize local flats independently of the presence or absence of plant cover. These subsurface-dwelling species may be protected from any superficial disturbance and, additionally, their sedentary habit may provide a rapid response to sediment disturbance. Most of these species are highly mobile in their search for food and selection of suitable habitat for burrowing (as in the case of predator polychaete *Glycinde multidens* and *Sthenolepis grubei*) or due to unpredictability in food availability (Pagliosa & Lana, 2005).

Temporal variations in population density and biomass of *Nereis oligohalina* were strongly correlated to seasonal variations in belowground biomass of *S. alterniflora*, suggesting that the reproductive strategies of the polychaete and the seasonal allocation of energy to belowground tissues of the marsh plant are synchronized (Pagliosa & Lana, 2000). Surface plant architecture played a major role in structuring epifaunal habitat specialist assemblages. Like submerged aquatic vegetation, the plant blades of *S. alterniflora* may enhance substratum availability and enable epiphytic development, attracting epifaunal grazing organisms and associated epibenthic predators (Bologna & Heck Jr., 1999; Edgar, 1999).

DETRITIVORES

The detritus food chain is the most import way to transfer matter in salt marshes and a large proportion of this detritus is not utilized on the marsh (Figure 4). Early estimates suggested that 45% of it is exported to the neighboring estuaries and flats (Little, 2000). Studies about the isotope ratios of the invertebrate concluded that movement and assimilation of carbon by invertebrates occurred at a scale of < 30 m (Guest & Connolly, 2004).

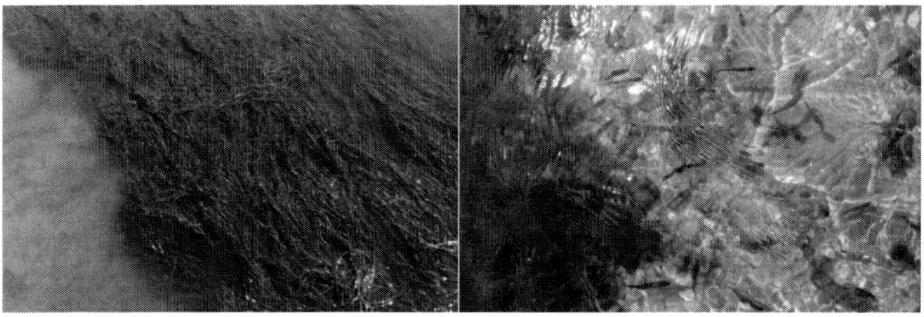

Figure 4. The debris exported by salt marshes helps fertilize neighboring areas, such as seagrass beds, which act as biodiversity hot spots and in which live several species of fishing importance.

The marsh periwinkle *Littoraria irrorata*, the salt marsh coffee-bean snail *Melampus bidentatus* and the talitrid amphipod *Uhlorchestia spartinophila* are common macroinvertebrates in salt marshes along the US Atlantic coast. They also represent potentially important links between microbial decomposer assemblages and higher trophic levels in the salt marsh system, because all three ingest detritus and, in turn, are consumed by mobile predators (Schindler et al., 1994).

Molluscs

Littoraria irrorata, a common species in east coast US, actually retreats upwards on *Spartina* stems as the water rises, so is seldom covered by the tide. When it is covered, it may become bloated because the incoming seawater often has a low salinity. It is also subject to devastating predation from the blue crab, *Callinectes sapidus*. Snails tethered on the marsh surface are five times more likely to die than those tethered high up on *Spartina* (Schindler et al., 1994; Little, 2000).

L. irrorata prefers and grows more rapidly on diets of dead cordgrass leaves than on either yellow-brown or live leaf-diets (Bärlocher & Newell, 1994). This preference is related to the fungal-biomass content of decomposing leaves, which are ingested at a higher rate than living leaf tissue. Newell and Bärlocher (1993) calculated that living fungal biomass constituted 25 to 30% of the ingestion of *L. irrorata* under laboratory conditions.

The intertidal distribution of *Melampus bidentatus* is more restricted than that of the larger marsh periwinkles and, in many salt marshes along the eastern coast of the USA, salt marsh coffee-bean snails tend to be found at relatively higher intertidal elevations (Daiber, 1982). Densities are typically < 700 individuals per m^2 (Peck et al., 1994). Like periwinkles, coffee-bean snails seem to prefer dead rather than living plant tissue and aged rather than unaged detritus (Rietsma et al., 1988). *Fundulus heteroclitus*, a very common and abundant resident fish species in tidal marshes along the east coast of North America, is among the known predators of *M. bidentatus* (Joyce & Weisberg, 1986).

Amphipods

The talitrid amphipod *Uhlorchestia spartinophila* is endemic to *Spartina alterniflora*-dominated marshes of the North American Atlantic coast (Bousfield & Heard, 1986), where it resides in spaces between overlapping leaf sheaths low down on the cordgrass shoots. Densities of 150 to 350 individuals per m^2 have been reported from marshes on Sapelo Island, Georgia (US), where these amphipods are broadly distributed through the intertidal zone wherever *S. alterniflora* is found (Covi & Kneib, 1995). Kneib et al., (1997) demonstrated that *U. spartinophila* grew and reproduced when fed solely on naturally decaying leaf parts of *S.*

alterniflora. This species, along with other common amphipods, constitutes an important part of the prey pool for higher-order consumer species in salt marshes (Covi & Kneib, 1995).

PREDATORS

Predation occurs at different levels (Figure 5). Although the best-known examples are blue crabs, some fish and alligators (mainly in North America), there are several organisms acting as secondary and tertiary consumers. In South America, some birds, such as the blue-black grass quit (*Volatinia jacarina*), the Brazilian canary (*Sicalis flaveola)*, and the great kiskadee (*Pitangus sulphuratus*), feed on insects. Small rodents (*Cavia aperea* and *Myocastor oypus*) feed on plant rhizomes and they can become prey to hawks and snakes (Seeliger et al., 2004).

Figure 5. Coastal birds search for shelter in the salt marshes. They are also important predators of fish and insects. The blue crab (*Callinectes sapidus*) is the main predator of invertebrates, such as molluscs.

Nifong & Silliman (2013) found evidence that apex predators can propagate non-consumptive interactions on their prey via perception of predatory risk-cues. They studied the impact of the alligator's predation in the salt marshes of the US. Blue crabs responded to alligator consumptive effects and predatory risk-cues by seeking refuge more often and by reducing foraging activity. In southeastern US salt marshes, alligators can occur in high densities during seasons of peak activity (April–November). In the case of blue crabs, two temporal constraints limit the use of salt marsh habitats: season and tidal stage (Fitz & Wiegert, 1991). Blue crabs can also become highly abundant during spring and summer months and often migrate offshore to deeper waters during winter months. Within creeks, alligators chiefly use marsh edge and benthic habitats to search and acquire prey items. These temporal and spatial constraints concentrate the window for the alligator–blue crab interactions on both a seasonal basis (spring and summer) and within marsh creeks, as opposed to higher elevation marsh platforms. Thus, the indirect effects of alligators are likely to be spatially confined in the marsh to the creek bank grasses, unless alligators can suppress crab densities so much that they do not enter marsh watersheds, a possibility that may not be likely given the lower energetic demands of a large-bodied ectotherm, such as the American alligator.

REFERENCES

Adam, P. (1990). *Saltmarsh Ecology*. Cambridge University Press, Cambridge.

Albano, M. J., Obenat, S., & Luppi, T. (2012). Community composition, assemblages and latitudinal pattern of benthic intertidal invertebrates at marshes in the Southwestern Atlantic Argentina Conferencia. In *CERF's Inaugural International Conference of the Americas* (Mar del Plata), 48-49.

Alberti, J., Daleo, P., Fanjul, E., Escapa, M., Botto, F., & Iribarne, O. (2015). Can a single species challenge paradigms of salt marsh functioning? *Estuaries and Coasts*, 38, 1178-1188.

Alberti, J., Daleo, P., Iribarne, O., Silliman, B. R., & Bertness, M. (2007). Local and geographic variation in grazing intensity by herbivorous crabs in SW Atlantic salt marshes. *Marine Ecology Progess Series*, 349, 235-243.

Bärlocher, F., & Newell, S. Y. (1994). Growth of the salt marsh periwinkle *Littoraria irrorata* on fungal and cordgrass diets. *Marine Biology*, 118, 109-114.

Bemvenuti, C. E. (1992). Benthic Invertebrates. In Seeliger, U. (editor). *Coastal plant communities of Latin America*. Academic Press, New York, New York, US.

Begon, M., Harper, J. L., & Towsend, C. R. (1996). *Ecology: individuals, populations and communities*. 3rd. ed. Blackwell Science, New York.

Bertness, M. D., & Hacker, S. D. (1994). Physical stress and positive associations among marsh plants. *American Naturalist*, 144, 363-372.

Boer, P. J. (1981). On the survival of populations in a heterogeneous and variable environment. *Oecologia*, 50, 39-53.

Bologna, P. A. X., & Heck Jr., K. L. (1999). Macrofaunal associations with seagrass epiphytes: relative importance of trophic and structural characteristics. *Journal of Experimental Marine Biology and Ecology*, 242, 21-39.

Bousfield, E. L., & Heard, R. W. (1986). Systematics, distributional ecology, and some host-parasite relationship of *Uhlorchestia auhleri* (Shoemaker) and *U.spartinophila*, new species (Crustacea: Amphipoda) endemic to salt marshes of the Atlantic Coast of North America. *Journal of Crustacean Biology*, 6, 264-274.

Canepuccia, A. D., Alberti, J., Pascual, J., Alvarez, G., Cebrian, J., & Iribarne, O. O. (2010). ENSO episodes modify plant/terrestrial – herbivore interactions in southwestern Atlantic salt marsh. *Journal of Experimental Marine Biology and Ecology*, 396, 42-47.

Chapman, V. J. (1974). *Salt marshes and salt deserts of the world*. Interscience Press, New York, New York, US.

Covi, M. P., & Kneib, R. T. (1995). Intertidal distribution, population dynamics and production of the amphipod *Uhlorchestia spartinophila* in a Georgia, USA, salt marsh. *Marine Biology*, 121, 447-455.

Costa, C. S. B., & Davy, A. J. (1992). Coastal salt marsh communities of Latin America. In Seeliger, U. (editor). *Coastal plant communities of Latin America*. Academic Press, New York, New York, US.

Daiber, F. C. (1982). *Animals of the tidal marsh*. Van Nostrand Reinhold, New York.

Daleo, P., & Iribarne, O. (2009). The burrowing crab *Neohelice granulata* affects the root strategies of the cordgrass *Spartina densiflora* in SW Atlantic salt marshes. *Journal of Experimental Marine Biology and Ecology*, 373, 66-71.

Edgar, G. J. (1999). Experimental analysis of structural versus trophic importance of seagrass beds. I. Effects on macrofaunal and meiofaunal invertebrates. *Vie et Milieu*, 49, 239-248.

Fitz, H. C., & Wiegert, R. G. (1991). Utilization of the intertidal zone of salt marsh by the blue crab *Callinectes sapidus*: density, return frequency, and feeding. *Marine Ecology Progress Series,* 76, 249-260.

Forattini, O. P., Rabello, E. X. & Pattoli, D. (1958). *Culicoides* da região Neotropical (Diptera: Ceratopogonidae). II – Observações sobre a biologia em condições naturais. *Arquivos da Faculdade de Higiene e Saúde Pública da Universidade de São Paulo*, 12, 1-52. [*Culicoides* of the Neotropical region (Diptera: Ceratopogonidae). II – Observations on biology under natural conditions. *Archives of the School of Hygiene and Public Health of the University of São Paulo*, 12, 1-52].

Furukawa, Y., Bentley, S. J., & Lavoie, D. L. (2001). Bioirrigation modeling in experimental benthic mesocosms. *Journal of Marine Research*, 59, 417-452.

Giere, O., & Pfannkuche, O. (1982). Biology and ecology of marine Oligochaeta, a review. *Oceanography and Marine Biology – An Annual Review*, 20, 173-308.

Gleason, M. L., Elmer, D. A., Pien, N. C., & Fisher, J. S. (1979). Effects of stem density upon sediment retention by salt marsh cordgrass. *Estuaries,* 2, 271-273.

Gleason, M. L., & Zieman, J. C. (1981). Influence of tidal inundation on internal oxygen supply of *Spartina alterniflora* and *Spartina patens*. *Estuarine, Coastal and Shelf Science*, 13, 47-57.

Graça, M. A., Newell, S. Y., & Kneib, R. T. (2000). Grazing rates of organic matter and living fungal biomass of decaying *Spartina alterniflora* by three species of salt marsh invertebrates. *Marine Biology*, 136, 281-289.

Guest, M. A., & Connolly, R. M. (2004). Fine-scale movement and assimilation of carbon in saltmarsh and mangrove habitat by resident animals. *Aquatic Ecology*, 38, 599-609.

Healy, B., & Walters, K. (1994). Oligochaeta in *Spartina* stems: the microdistribution of Enchytraeidae and Tubificidae in a salt marsh, Sapelo Island, USA. *Hydrobiologia*, 278, 111-123.

Hobach, W. W., Svatos, T. M., Spomer, S. M. & Higley, L. G. (1999). Trap colour and placement effects estimates of insects family level and diversity in a Nebraska Salt Marsh. *Entomologia Experimentalis et Applicata*, 91, 393-402.

Iribarne, O., Martinetto, P., Schwindt, E., Botto, F., Bortolus, A., & García Borboroglu, P. (2003). Evidence of habitat displacement between two common soft-bottom SW Atlantic intertidal crabs. *Journal of Experimental Marine Biology and Ecology*, 296, 167-182.

Iribarne, O., Bortolus, A., & Botto, F. (1997). Between-habitat differences in burrow characteristics and trophic modes in the southwestern Atlantic burrowing crab *Chasmagnathus granulata*. *Marine Ecology Progress Series*, 155, 137-145.

Jackson, D., Mason, C. F., & Long, S. P. (1985). Macroinvertebrate populations and production on a salt marsh in east England dominated by *Spartina anglica*. *Oecologia*, 65, 406-411.

Joyce, A. A., & Weisberg, S. B. (1986). The effects of predation by the mummichog, *Fundulus heteroclitus* (L.), on the abundance and distribution of the salt marsh snail, *Melampus bidentatus* (Say). *Journal of Experimental Marine Biology and Ecology*, 100, 295-306.

Kline, D. L., & Axtell, R. C. (1976). Salt marshes *Culicoides* (Diptera: Ceratopogonidae): species, seasonal abundance and comparisons of trapping methods). *Mosquito News*, 36, 1-10.

Kneib, R. T. (1997). The role of tidal marshes in the ecology of estuarine nekton. *Oceanography and Marine Biology – An Annual Review*, 35, 163-220.

Kneib, R. T., Newell, S. Y., & Hermeno, E. T. (1997). Survival, growth and reproduction of the salt marsh amphipod *Uhlorchestia spartinophila* reared on natural diets of senescente and dead *Spartina alterniflora* leaves. *Marine Biology*, 128, 423-431.

Kuijper, D. P. J., Nijhoff, D. J., & Bakker, J. P. (2004). Herbivory and competition slow down invasion of a tall grass along a productivity gradient. *Oecologia,* 141, 452-459.

Lana, P. C., & Guiss, C. (1992). Macrofauna – plant – biomass interactions in an euhaline salt marsh in Paranagua Bay (SE Brazil). *Marine Ecology Progress Series*, 80, 57-64.

Levin, L. A., & Talley, T. S. (2002). Natural and manipulated sources of heterogeneity controlling early faunal development of a salt marsh. *Ecological Applications*, 12, 1785-1802.

Little, C. (2000). *The biology of soft shores and estuaries*. Oxford University Press. Oxford.

Louda, S. M., & Rodman, J. E. (1996). Insect herbivory as a major factor in the shade distribution of a native crucifer (*Cardamine cordifolia* A. Gray, Bittercress). *Journal of Ecology*, 84, 229-237.

Martinetto, P., Montemayor, D. I., Alberti, J., Costa, C. S. B., & Iribarne, O. (2016). Crab bioturbation and herbivory may account for variability in carbon sequestration and stocks in South West Atlantic Salt Marshes. *Frontiers in Marine Science*, 3, 1-12.

Martinetto, P., Palomo, G., Bruschetti, M., & Iribarne, O. (2011). Similar effects on sediments structure and in faunal community of two competitive intertidal soft-bottom burrowing crab species. *Journal of Marine Biological Association of United Kingdom*, 91, 1385-1393.

Martinetto, P., Valiñas, M., Palomo, G., & Iribarne, O. (2007). Negative interactions between two SW Atlantic intertidal crabs in soft bottom habitats. *Marine Biology*, 151, 1479-1490.

Merritt, R. W., Cummins, K. W. & Berg, M. B. (2008). *An introduction to the Aquatic Insects of North America*. 4th ed. Kendall/Hunt Publishing Company, Dubuque.

Newell, S. Y., & Bärlocher, F. (1993). Removal of fungal and total organic matter from decaying cordgrass leaves by shredder snails. *Journal of Experimental Marine Biology and Ecology*, 171, 39-49.

Nifong, J. C., & Silliman, B. R. (2013). Impacts of a large-bodied, apex predator (*Alligator mississippiensis* Daudin 1801) on salt marsh food webs. *Journal of Experimental Marine Biology and Ecology,* 440, 185-191.

Oremland, R. S., Marsh, L. M., & Polcin, S. (1982). Methane production and simultaneous sulphate reduction in anoxic, salt marsh sediments. *Nature*, 296, 143-145.

Pagliosa, P. R., & Lana, P. C. (2000). Population dynamics and secondary production of *Nereis oligohalina* (Nereididae: Polychaeta) from a subtropical marsh in a southeast Brazil. *Bulletin of Marine Science*, 67, 259-268.

Pagliosa, P. R., & Lana, P. C. (2005). Impact of plant cover removal on macrobenthic community structure of a subtidal salt marsh. *Bulletin of Marine Science*, 77, 1-17.

Pascual, J., Canepuccia, A. D., Alberti, J., Daleo, P., & Iribarne, O. (2015). Rainfall intensity modulates the interaction between the marsh cordgrass. *Spartina densiflora* and the mouse *Akodon azarae*. *Marine Ecology Progress Series*, 523, 71-80.

Peck, M. A., Fell, P E., Allen, E. A., Gieg, J. A., Guthke, C, R., & NewKirk, M. D. (1994). Evaluation of tidal marsh restoration: comparison of selected macroinvertebrate populations on the restored impounded valley marsh and an unimpounded valley marsh within the same salt marsh system in Connecticut, US. *Environmental Management*, 18, 283-293.

Peterson, B. J., & Howarth, R. W. (1987). Sulfur, carbono, and nitrogen isotopes used to trace organic matter flow in the saltmarsh estuaries of Sapelo Island, Georgia. *Limnology and Oceanography*, 32, 1195-1213.

Pennings, S. C., & Bertness, M. D. (2001). Salt marsh communities. In Bertness, M. D. and Hay, M. E. (editors). *Marine community ecology*. Sinauer, Sunderland, Massachusetts, US.

Pupedis, R. J. (1997). Aquatic insects of the west-river and salt marshes of Connecticut. *Yale Forest and Environmental Studies Bulletin*, 100, 162-177.

Rietmsma, C. S., Valiela, I., & Buchsbaum, R. (1988). Detrital chemistry, growth and food choice in the salt marsh (*Melampus bidentatus*). *Ecology*, 69, 261-266.

Rand, T. A. (2002). Variation in insect herbivory across a salt marsh tidal gradient influences plant survival and distribution. *Oecologia*, 132, 549-558.

Rosenberg, D. M., Danks, H. V., & Lehmkuhl, D. M. (1986). Importance of insects in environmental impact assessment. *Environmental Management*, 10: 773-783.

Rozas, L. P., & LaSalle, M. W. (1990). A comparison of the diets of the Gulk killifish, *Fundulus grandis*, entering and leaving a Mississippi brackish marsh. *Estuaries*, 13, 332-336.

Sacco, J. N., Seneca, E. D., & Wentworth, T. R. (1994). Infaunal Community development of artificially established salt marshes in North Carolina. *Estuaries*, 17, 489-500.

Schindler, D. E., Johnson, B. M., MacKay, N. A., Bouwes, N., & Kitchell, J. F. (1994). Crab: snail size-structured interactions and salt marsh predation gradients. *Oecologia*, 97, 49-61.

Schrama, M., Berg, M., & Olff, H. (2012). Ecosystem assembly rules: the interplay of green and brown webs during salt marsh succession. *Ecology*, 93, 2353-2364.

Seeliger, U., Cordazzo, C., & Barcellos, L. (2004). *Areias do Albardão*. Ecoscientia, Rio Grande, Brazil. [*Sands of Albardão*].

Seliskar, D. M., Gallagher, J. L., Burdick, D. M., & Mutz, L. A. (2002). The regulation of ecosystem functions by ecotypic variation in the dominanr plant: a *Spartina alterniflora* saltmarsh case study. *Journal of Ecology*, 90, 1-11.

Silliman, B. R., & Bertness, M. D. (2002). A trophic cascade regulates salt marsh primary production. *Proceedings of the National Academy of Sciences*, 99, 10500-10505.

Spinelli, G. R. & Wirth, W. W. (1993). Los Ceratopogonidae de la Argentina (Insecta: Diptera), p. 1-124. In: Castellano, Z. (ed.). *Fauna de Agua Dulce de la República Argentina*. Estudio Sigma, Buenos Aires. [The Ceratopogonidae of Argentina (Insecta: Diptera), p. 1-124. In: Castellano, Z. (ed.). *Freshwater fauna of the Argentine Republic*. Estudio Sigma, Buenos Aires].

Teal, J. M., & Kanwisher, J. (1961) Gas exchange in a Georgia salt marsh. *Limnology and Oceanography*, 6, 388-399.

Teal, J. M., & Kanwisher, J. (1966). Gas transport in the marsh grass *Spartina alterniflora*. *Journal of Experimental Botany*, 17, 355-361.

Warren, R. S., & Neiring, W. A. (1993). Vegetation change on a northeast tidal marsh: interaction of sea-level rise and marsh accretion. *Ecology*, 74, 96-103.

Whitcraft, C. R., & Levin, L. A. (2007). Regulation of benthic algal and animal communities by salt marsh plants: impact of shading. *Ecology*, 88, 904-917.

Wirth, W. W., & Stone, A. (1968). Aquatic Diptera. In Usinger, R. L. (editor). *Aquatic Insects of California*. University of California Press, Berkeley, pp. 372-482.

Wyatt, T. D. (1986). How a subsocial intertidal beetle, *Bledius spectabilis*, prevents flooding and anoxia in its burrow. *Behavioral Ecology and Sociobiology*, 19, 323-331.

INDEX

E

F

G

H

I

L

M

N

O

P

R

S

T

U

V

W